THE GLOBAL FOOD CRISIS

BOOK SOLD
PROPERTY

RICHMOND HILL
PUBLIC LIBRARY

JAN 6 2010

RICHMOND GREEN
905-780-0711

D0910843

Studies in International Governance is a research and policy analysis series from the Centre for International Governance Innovation (CIGI) and Wilfrid Laurier University Press. Titles in the series provide timely consideration of emerging trends and current challenges in the broad field of international governance. Representing diverse perspectives on important global issues, the series will be of interest to students and academics while serving also as a reference tool for policy-makers and experts engaged in policy discussion. To reach the greatest possible audience and ultimately shape the policy dialogue, each volume will be made available both in print through WLU Press and, twelve months after publication, free online under the Creative Commons License.

CIGI

The Centre for International
Governance Innovation
Centre pour l'innovation dans
la gouvernance internationale

THE GLOBAL FOOD CRISIS
Governance Challenges and Opportunities

Jennifer Clapp and
Marc J. Cohen, editors

Wilfrid Laurier University Press

WLU

Wilfrid Laurier University Press acknowledges the financial support of the Government of Canada through its Book Publishing Industry Development Program for its publishing activities. Wilfrid Laurier University Press acknowledges the financial support of the Centre for International Governance Innovation. The Center for International Governance Innovation gratefully acknowledges support for its work program from the Government of Canada and the Government of Ontario.

Library and Archives Canada Cataloguing in Publication

The global food crisis : governance challenges and opportunities / edited by Jennifer Clapp and Marc J. Cohen.

(Studies in international governance series)
Co-published by the Centre for International Governance Innovation.
Includes bibliographical references and index.
Issued also in electronic format.
ISBN 978-1-55458-192-4

1. Food prices—Developing countries. 2. Food supply—Developing countries. 3. Food relief— International cooperation. 4. Crops and climate—Developing countries. 5. Sustainable agriculture. 6. Agricultural systems. I. Clapp, Jennifer, 1963– II. Cohen, Marc J., 1952– III. Centre for International Governance Innovation IV. Series: Studies in international governance

HD9000.5.G5826 2009   338.1'91724   C2009-903706-8

RICHMOND
PUBLIC LIBRARY

JAN 6 2010

RICHMOND GREEN
905-780-0711

Library and Archives Canada Cataloguing in Publication

The global food crisis [electronic resource] : governance challenges and opportunities / edited by Jennifer Clapp and Marc J. Cohen.

(Studies in international governance series)
Co-published by the Centre for International Governance Innovation.
Includes bibliographical references and index.
Electronic edited collection in PDF, ePub, and XML formats.
Issued also in print format.
ISBN 978-1-55458-198-6

1. Food prices—Developing countries. 2. Food supply—Developing countries. 3. Food relief— International cooperation. 4. Crops and climate—Developing countries. 5. Sustainable agriculture. 6. Agricultural systems. I. Clapp, Jennifer, 1963– II. Cohen, Marc J., 1952– III. Centre for International Governance Innovation IV. Series: Studies in international governance

HD9000.5.G5826 2009a   338.1'91724

Cover design by David Drummond. Text design by Catharine Bonas-Taylor.

© 2009 The Centre for International Governance Innovation (CIGI) and Wilfrid Laurier University Press

For permission to reprint the articles "Impact of Climate Change and Bioenergy on Nutrition," by Marc Cohen, Cristina Tirado, Noora-Lisa Aberman (IFPRI), and Brian Thompson (FAO), and "A Stronger Global Architecture for Food and Agriculture: Some Lessons from FAO's History and Recent Evaluation," by Daniel J. Gustafson and John Markie, the publishers acknowledge the Food and Agriculture Organization of the United Nations and the International Good Policy Research Institute. Both articles are copyright IFRPI and FAO, 2008.

This book is printed on FSC recycled paper and is certified Ecologo. It is made from 100% post-consumer fibre, processed chlorine free, and manufactured using biogas energy.

Printed in Canada

Every reasonable effort has been made to acquire permission for copyright material used in this text, and to acknowledge all such indebtedness accurately. Any errors and omissions called to the publisher's attention will be corrected in future printings.

No part of this publication may be reproduced, stored in a retrieval system or transmitted, in any form or by any means, without the prior written consent of the publisher or a licence from The Canadian Copyright Licensing Agency (Access Copyright). For an Access Copyright licence, visit www.accesscopyright.ca or call toll free to 1-800-893-5777

CONTENTS

Introduction

Part 1: The Causal Factors behind the Food Crisis

Part 2: Immediate Governance Challenges and Proposals: Food Aid, Trade Measures, and International Grain Reserves

Part 3: Longer-Term Ecological Concerns
and Governance Responses

Part 4: Strategies to Promote Food Security and
Sustainable Agriculture: The Way Ahead

LIST OF
ILLUSTRATIONS

LIST OF TABLES AND BOXES

Tables

Boxes

FOREWORD

R apidly increasing international food prices during 2007 and the first half
of 2008 attracted much attention from international institutions and policy-
makers, analysts, and news media around the world. Complacency caused by
falling real food prices during the period 1974–2000, and an attitude in most
governments that agriculture was yesterday's priority, was suddenly replaced
by great concern about the availability of food currently and in the future.

As street demonstrations and riots, justified by the rising food prices, provided opportunities to air other grievances about social injustice in an increasing number of countries, policy-makers went into panic mode. Street riots,
particularly when driven by dramatic deteriorations in the access to food or
other basic necessities, are very newsworthy, and the news media in their zeal
contributed to the nervousness of politicians and the public they serve. Dramatic decreases in food prices since mid-2008 attracted virtually no media
attention but may unfortunately lead to renewed and totally misplaced complacency among policy-makers.

As national governments scrambled to slow the tide of dissatisfaction in
urban areas and compensate politically active consumers for losses in their purchasing power caused by higher food prices, the international consequences of
their policy measures were all but ignored. In order to control domestic rice
prices, some traditional rice exporters, such as India, introduced draconian
export restrictions that contributed to a tripling of international rice prices in a
matter of months. A major soybean and maize exporter, Argentina, attempted
to expand export taxes and thereby reduced production incentives and international supplies, while the United States expanded subsidies for maize production for biofuel, a policy that contributed to food-price increases in the first
place. A variety of other policy measures with international consequences were
introduced around the world. Fortunately, two major grain exporters, Thailand
and the United States, did not introduce export restrictions and thus helped to
avoid even larger international price increases.

As illustrated by events that took place during the international food-price run-up, food systems around the world are increasingly integrated into a global system—a system that requires international governance. Unfortunately, as stated by Clapp and Cohen in chapter 1 of this book, the "global governance of food and agriculture is fragmented and incoherent." With that as the postulate, one with which I agree, Clapp and Cohen invited a group of outstanding policy analysts, economists, and others with expertise in food policy to enhance the understanding of the food crisis, identify its causes, determine how it compares with past food crises, show how national policies link to the global economic context, explain the role of the private sector, and identify which immediate and longer-term governance challenges and strategies might strengthen the global governance of the food system. This book, which I found to be most informative and constructive, is a result of their deliberations.

The main strengths of the book are its analytical description of the causes of the food crisis and its presentation of a clear, evidence-based set of strategies and proposals for strengthening national and international governance of the global and national food systems. The choice of authors, who represent a variety of perspectives in their analyses, is an important added attraction of the book. The book provides much new relevant evidence for students of food policy and global governance as well as for policy analysts, advisors, and policy-makers.

Per Pinstrup-Andersen
H.E. Babcock Professor of Food, Nutrition and Public Policy, Cornell University; J. Thomas Clark Professor of Entrepreneurship, Cornell University; Professor of Applied Economics and Management, Cornell University; Professor of Agricultural Economics, Copenhagen University

ACKNOWLEDGEMENTS

This book originated with a workshop held in Waterloo, Ontario, in December 2008 on the theme of international governance responses to the food crisis. We would like to thank the Centre for International Governance Innovation for financial support for that workshop and for its support for the publication of this volume, as well as the International Food Policy Research Institute for providing organizational support for the workshop. Special thanks go to David Norris, Ryan Pollice, and Linda Swanston for providing outstanding editorial assistance in the preparation of the manuscript. We are very grateful to number of additional people who have assisted in the preparation of both the workshop and this volume: Max Brem, Kimberly Burnett, Patrick Clark, Briton Dowhaniuk, Jessica Hanson, Clare Hitchens, Jennifer Jones, Rob Kohlmeier, Joshua Lovell, and Leslie Macredie. We would also like to express our gratitude to the participants in our workshop and in particular to the contributors to this book for working with us on a tight schedule. Finally, we owe a debt of thanks to our families and friends, for their patience and support throughout the publication process.

ACRONYMS AND ABBREVIATIONS

ACIA   Arctic Climate Impact Assessment
ADB   Asian Development Bank
ADMARC   Agricultural Development and Marketing Corporation
AGOA   African Growth and Opportunity Act (United States)
AGRA   Alliance for a Green Revolution in Africa
AICD   African Infrastructure Country Diagnostic
AIDS   Acquired Immune Deficiency Syndrome
AKST   Agricultural Knowledge, Science and Technology
CAADP   Comprehensive African Agricultural Development Pro-
            gramme
CARD   Center for Agricultural and Rural Development
CGIAR   Consultative Group on International Agricultural Research
CIGI   Centre for International Governance Innovation
CO$_2$ Carbon Dioxide
COMESA   Common Market for Eastern and Southern Africa
CRP   Conservation Reserve Program
CSSD   Consultative Subcommittee on Surplus Disposal
CTA   Comparative Technology Assessment
DAC   Development Assistance Committee
ECHO   European Community Humanitarian Aid Department
ECOSOC   UN Economic and Social Council
ECOWAS   Economic Community of West African States
EPA   Environmental Protection Agency (US)
EU   European Union
FAC   Food Aid Convention
FAO   Food and Agriculture Organization of the United Nations
G77   Group of 77 Developing Countries
G8   Group of Eight
G20   Group of Twenty
GATT   General Agreement on Tariffs and Trade

GFCRP   Global Food Crisis Response Program
GHG   Greenhouse Gas
GMB   Grain Marketing Board (Zimbabwe)
HIPC   Heavily Indebted Poor Countries
HIV   Human Immunodeficiency Virus
HRCR   Human Rights Council Report (UN)
IAASTD   International Assessment of Agricultural Knowledge,
             Science and Technology for Development
IATP   Institute for Agriculture and Trade Policy
IDRC   International Development Research Centre (Canada)
IEA   International Energy Agency
IEE   Independent External Evaluation (of the FAO)
IFAD   International Fund for Agriculture and Development
IFI   International Financial Institution
IFPRI   International Food Policy Research Institute
IMF   International Monetary Fund
IPA   Immediate Plan of Action
IPCC   Intergovernmental Panel on Climate Change
IPR   Intellectual Property Right
LDC   Least Developed Countries
LEAD   Livestock, Environment, and Development
LIFDC   Low Income Food Deficit Country
M bpd   Million barrels per day
MCC   Millennium Challenge Corporation
MDG   Millennium Development Goal
MENA   Middle East and North Africa Region
NATO   North Atlantic Treaty Organization
NEPAD   New Partnership for Africa's Development
NGO   Nongovernmental Organization
OCHA   Office for the Co-ordination of Humanitarian Affairs
ODA   Official Development Assistance
OECD   Organization for Economic Co-operation and Development
PEPFAR   President's Emergency Plan for AIDS Relief (United
             States)
PL 480   Public Law 480 (US food aid program)
SMA   Seed Maize Association (Zimbabwe)
TAFAD   Trans-Atlantic Food Assistance Dialogue
TNC   Transnational Corporation
UA   Urban Agriculture
UN   United Nations

UNCTAD	United Nations Conference on Trade and Development
UNDP	United Nations Development Programme
UNEP	United Nations Environment Programme
UNICEF	United Nations Children's Fund
USAID	US Agency for International Development
USDA	United States Department of Agriculture
USTR	United States Trade Representative
WFC	World Food Council
WFP	World Food Programme
WHO	World Health Organization
WTO	World Trade Organization

CHAPTER 1

The Food Crisis and Global Governance

Jennifer Clapp and Marc J. Cohen

The rapid and sharp rises in food prices in late 2007 and early 2008 were a stark reminder of the fragility and volatility of the global food system. As food prices shot to dizzying heights, the world's poor people—those most vulnerable to food price rises—were hard hit. Over 850 million people were already considered food inse-cure when the prices began to rise, and the situation pushed more into that category. Although many of the affected people are smallholder farmers, they are also fre-quently net purchasers of food, and so the price increases had severe impacts. Civil unrest and "food riots" erupted in over forty developing countries as people's ability to command food suddenly dropped. While food prices on international markets eased considerably by the fall of 2008, they still remained some 30 percent above 2005 levels. Domestic food prices in developing countries have not fallen back and by mid-2009 were higher still than levels seen in early 2008. By early 2009, the Food and Agriculture Organization of the United Nations (FAO) announced that the number of chronically hungry people in the world had climbed to over 1 billion. When making this announcement, Jacques Diouf, director-general of the FAO, stated bluntly in an interview with the *Financial Times*, "The food crisis is not over" (quoted in Blas 2009).

This book provides a set of analyses of the global food crisis with a partic-ular focus on the challenges and opportunities it presents for the governance of the international food and agriculture system. Its aim is to provide a snapshot of the range of debate over the causes of the crisis, its consequences in both the short and long term, and proposals for the way forward. The chapters are based on presentations originally given at a December 2008 workshop held at the Centre for International Governance Innovation (CIGI) in Waterloo, Ontario.

They represent a range of viewpoints on the effectiveness of existing governance institutions and processes for the global food and agriculture system, their role in precipitating and/or alleviating the crisis, and the reforms that might be required to more effectively deliver sustainable food security for all.

This introductory chapter first provides a brief overview of the extent of the food crisis. It then examines the key themes of the book, including debates over the causes of the crisis, the implications for short-term international governance responses, the ecological dimensions of the crisis, and the considerations for formulating longer-term governance responses that would alleviate the crisis and support a more sustainable and resilient global food system.

Snapshot of the Crisis

Beginning in late 2006, food prices began to rise, marking a change from the trend of the previous four decades of slow but steady declines in agricultural commodity prices (albeit with some sharp upward spikes along the way). Prices continued their ascent in 2007, and by early 2008 began to shoot up sharply, particularly for some key food staple commodities such as rice, wheat, and maize. According to the FAO, and as shown in Figure 1.1, the index of nominal food prices more than doubled between 2002 and mid-2008 (see FAO 2008).

Figure 1.1
FAO Food Price Index, 2005–2009

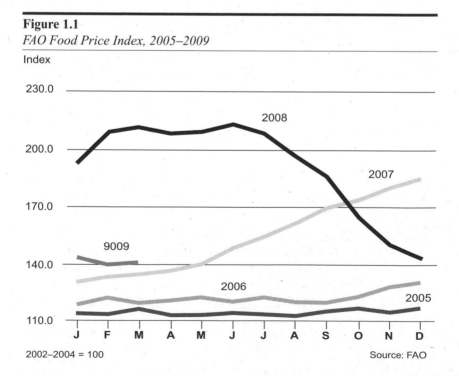

2002–2004 = 100   Source: FAO

Rapidly rising food prices are a problem because poor people in developing countries often spend upward of 60–80 percent of their income on food. When food prices double, people suddenly can command only half of the food that they once could, and poor consumers frequently make adjustments in what they consume, cutting out nutritious fruits, vegetables, and animal source foods so that they can maintain consumption of high-energy staples. Moreover, when people are already living in or on the edge of poverty, sudden changes in their ability to command food are destabilizing. The food price riots of 2007 and 2008 are not at all surprising when seen in this light. Although global grain production increased in 2008, and international food prices dropped considerably after record heights (just as the global economy began a major slowdown), food prices in mid-2009 still remain significantly above 2005 levels.

Within developing countries, meanwhile, domestic prices remain high even as international prices have fallen. In some cases, local food prices have risen to levels higher than those seen in 2008. For developing countries dependent on food imports, the global economic recession that took hold in mid-2008 has meant fewer resources available to finance food imports. Tight supplies have led to rising prices in these countries, with devastating impacts on poor people.

In the early 1990s, 20 percent of the developing world's population was chronically undernourished. One of the key Millennium Development Goal (MDG) targets set by the global community at the 2000 UN Millennium Summit was to cut that figure to 10 percent by 2015. Although progress was slow, by the 2003-2005 period the proportion in extreme hunger had fallen to 16 percent. But the food price rises of 2007–08 have driven that figure back up to nearly 18 percent (FAO 2008). The World Food Programme (WFP) estimates that higher food prices will set back any progress on the MDGs by at least seven years (WFP 2009). Moreover, in absolute terms, the *number* of hungry people has been on the rise since 1995 (see Figures 1.2 and 1.3).

After several tumultuous years on international food markets, there is widespread concern about food price volatility and its broader impacts. While the sharp price rises and falls are problematic in themselves, the situation has cast light on other aspects of the global food system that are in need of repair, as outlined below.

Investment in Agriculture

Investment in agriculture, especially in developing countries, has fallen since the 1980s, affecting production, infrastructure, and institutional capacity, as well as the ability of farmers to respond to rising prices with increased production. This underinvestment in developing country agriculture has been practiced by both poor country governments and international donor agencies. Agricultural investment in 2007 was only 4 percent of public spending in sub-Saharan Africa, while

Figure 1.2
Proportion of Undernourished People in the Developing World

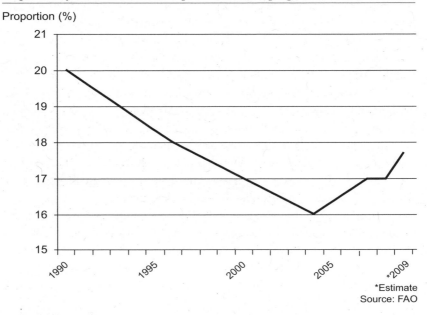

Proportion (%)

*Estimate
Source: FAO

Figure 1.3
Number of Undernourished People in the Developing World

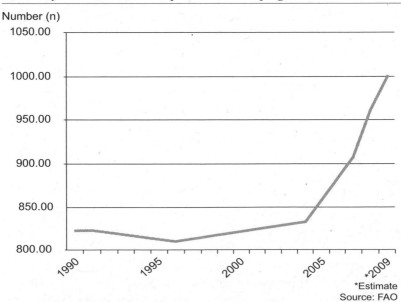

Number (n)

*Estimate
Source: FAO

the share of agriculture in official development assistance declined from 18 percent in 1979 to just 3.5 percent in 2004 (World Bank 2007, 40–42). The World
Bank's lending for agriculture dropped even more sharply, from 30 percent of
its loans in 1980 to just 3 percent in recent years (World Bank 2008). Donor
support for public agricultural research has barely kept pace with inflation.

Unequal Trading System

The international agricultural trade system is widely seen to be highly unequal,
which is why the Doha Round of World Trade Organization (WTO) talks have
focused so closely on rectifying the imbalances in the agricultural sector. Rich
industrialized countries currently spend over US$300 billion per year on agri-
cultural subsidies to support their agricultural production and trade. Developing country governments cannot afford to provide a comparable level of
subsidies, and even if they could, donors have made a more market-oriented
approach to agriculture a condition of assistance for the past three decades.
This situation has led to what many see as a highly uneven playing field, which
has dampened incentives for agricultural production in developing countries.

Food Insecurity

Large numbers of low-income people, especially in the developing world, are
food insecure. Although many of the world's poor and food insecure live in
rural areas with livelihoods closely tied to the agricultural sector, rapidly rising prices do not necessarily lead to rising incomes. Rising production costs,
weak infrastructure, and lack of credit have all hampered the ability of the
world's poor to respond to food price rises with greater production. Conflict,
poor weather conditions, and growing numbers pushed into extreme poverty
since 2008 (a result of the global economic crisis) have also meant that more
people are in need of food aid to meet their daily caloric requirements. Yet in
eras of economic uncertainty, levels of international food aid tend to fall just
as need rises. Moreover, the current international food aid system has been critiqued as being highly inefficient.

Climate Change

Global agricultural production is also threatened by ecological degradation and
climate change. As the climate warms and water is increasingly in short supply, it has become apparent that the way in which the bulk of modern agriculture is pursued—with large-scale farms for both crops and livestock that are
heavily reliant on water and fossil-fuel-based inputs—is not sustainable. Already

the impacts of climate change and water shortages are causing a drag on increases in productivity, leading many to call for the adoption of more sustainable agricultural methods and systems.

Implications for Governance

The governance framework for the global food and agricultural system is in many ways based on past conditions, practices, and understandings of how best to promote global food security. The volatility and the vulnerability highlighted by the current food crisis demands a closer look at the international food and agriculture governance architecture—to examine both its role in the current crisis, and the potential for improvements.

As it stands, global governance of food and agriculture is fragmented and incoherent. Many international institutions claim a role, mandates overlap, and power structures within the relevant institutions vary considerably. The Bretton Woods institutions (the World Bank and the International Monetary Fund) play a major role in financing agricultural and rural development projects and programs, and food imports. They are strongly market-oriented, with the principal donor-country governments holding decisive voting power on the boards of directors. The Rome-based UN food and agriculture agencies (FAO, WFP, International Fund for Agricultural Development), in contrast, have more balanced North–South representation on their governing bodies and are significant players in norm setting, data collection, technical assistance, and emergency aid. FAO and UN human rights institutions (the Office of the High Commissioner, the Human Rights Council's Special Rapporteur on the Right to Food, and the Committee on Economic, Social, and Cultural Rights) temper neo-liberal approaches to food and agricultural development with an emphasis on a rights-based orientation that makes access to food for all the touchstone for policy analysis. Within this divided governance framework, debate continues about the appropriate roles of state, market, and civil society in achieving food security. There is evidence that power may be shifting. The wealthy governments of the North, acting bilaterally and through "donor clubs" such as the Organization for Economic Co-operation and Development and the G8, continue to provide the vast bulk of official development assistance, but new donors are emerging among governments of oil-exporting countries and higher-income developing countries. Similarly, power within the WTO has shifted from Northern dominance to a contentious stalemate between North and South. What this means for the future is still playing out, and the food crisis will have a major bearing on the end result.

Meanwhile, private actors loom large in the global food system, with farm input sales, output marketing, and food retailing all increasingly concentrated in the hands of a few large transnational companies. In light of the problematic global governance structures, such matters as competition policy, food safety, public health, and poverty reduction are not well coordinated on a world scale. Regulatory capacity has not kept pace with the global integration of markets for goods, services, and money.

Map of This Book

The contributors to this volume present a range of perspectives. As either policy-makers or academics, each author brings their own take to the broad questions that frame this book. The aim was not to bring together a collection of like-minded analysts but rather to present a genuine debate on the future of global food security. The chapters in this book thus reflect a variety of viewpoints.

The first part of the book examines the causal factors behind the food crisis. Although there is a widely agreed range of possible triggers to the crisis, the relative weight of each is contested. In Chapter 2, Anuradha Mittal provides an assessment of the factors that have been identified as key contributors to food price rises. She argues that the longer-term structural causes are as important as the more immediate short-term causes, and that governance responses must make genuine progress toward reducing the inequalities between rich and poor countries if we are to adequately promote food security for all. In Chapter 3, Sue Horton looks at the analysis of the causes of the current crisis though a historical lens by drawing parallels with the 1970s food crisis. She argues that the current crisis is in many ways similar to the 1970s episode of rapid and steep food price rises, in particular in terms of both the immediate triggers and the underlying structural inequities. In making this comparison, she asks whether we have learned the right lessons from the past.

In Chapter 4, Jennifer Clapp focuses on the role of global macroeconomic factors in precipitating the food price rises. She argues that these factors, in particular financial speculation on agricultural commodity markets, have been largely underplayed in official policy responses to the crisis. In Chapter 5, Kim Elliott highlights the important role of US biofuel policy in precipitating the crisis. She argues that demand for biofuels is directly linked to the rise in both corn and soy prices and to some extent higher prices for other food crops. At the same time, however, she notes that while biofuel demand has wreaked havoc on food prices, the rise in production of corn-based ethanol has not contributed significantly to energy security or environmental goals.

The second part of the book looks more closely at the immediate governance challenges posed by rising food prices, in particular, emergency response measures. In Chapter 6, Raymond Hopkins makes the case for basing international food aid policy on insurance principles rather than the current system in which food aid levels are pro-cyclical, with emergency assistance reactive and completely dependent on the willingness of donors to respond. An insurance-based system would ensure that resource flows are stabilized in a way that would improve timely access to food, particularly in times of need. In Chapter 7, Stuart Clark makes the case for reform of the current Food Aid Convention. He argues that there is considerable scope to broaden the convention to make it more flexible and efficient, and that there is a need to think beyond food aid to "food assistance." Such reforms, he argues, would provide more reliable and meaningful food assistance.

Gawain Kripke argues in Chapter 8 that the world's largest food aid donor, the United States, must undertake long-overdue reforms to end the requirement that all of its food aid be in the form of US-grown agricultural commodities. Allowing for aid in the form of cash to enable local and regional purchase of food in developing countries themselves would, he argues, be much more efficient and would also support local agricultural development in the world's poorest countries. In Chapter 9, Frederic Mousseau provides a critique of current proposals for both real and virtual international food reserves. He argues instead in favour of nationally and regionally held food stocks in developing countries as being the most fruitful way to provide a safety net for the world's hungry.

The third part of the book looks at the longer-term ecological concerns associated with the current global food and agriculture system that have been highlighted by the food crisis and examines potential governance responses. In Chapter 10, Cristina Tirado, Marc J. Cohen, Noora-Lisa Aberman, and Brian Thompson discuss the challenges that climate change places on nutrition, particularly in the developing world. They argue that agriculture can play an important role in both climate change adaptation and mitigation strategies. They further argue that a human rights-based approach enables an embrace of environmental concerns when addressing both climate and nutritional issues. In Chapter 11, Tony Weis warns of the ecological and social problems associated with industrial agriculture, which is highly dependent on fossil fuels. He argues that the recent volatility in the food system is linked to the rise of industrial agriculture and that as this becomes increasingly recognized, there is also growing opportunity for reform. In Chapter 12, Noah Zerbe discusses the debate over agricultural biotechnology—one of the "magic bullets" often presented as a way around the ecological constraints. He contends that we should not necessarily see agricultural biotechnology as a solution to the crisis, largely because the food crisis is not a crisis of production but rather is linked to the marketization of food security. In this sense, he argues, solutions to the crisis must include efforts to re-embed food security in social relations rather than solely rely on production-enhancing technology.

The fourth and final part of the book examines strategies to promote food security and sustainable agriculture in today's global context. In Chapter 13, Daniel Gustafson and John Markie look back at the dilemmas faced more than half a century ago by the founders of the international food governance system and see striking parallels with today. They argue that buy-in from national governments and coordination with a wide variety of international organizations that address food and agriculture are essential for building a more effective international food governance system. In Chapter 14, Emmy Simmons and Julie Howard take a look specifically at the challenges faced by sub-Saharan African countries in the context of the food crisis, as well as at the role donors, particularly the United States, play in meeting those challenges. They argue that the United States must move from primarily a food aid response to food crises in Africa to broader support for agricultural development on the continent. This broader assistance, they argue, must embrace—and be built upon—African governments' own agenda for agriculture. In Chapter 15, Mark Redwood looks at the potential role of urban agriculture in mitigating hunger in the face of the current food crisis. For much of the past thirty years, rural areas have been the most food insecure, but recent price surges have precipitated a rise in urban hunger. He concludes that the promotion of urban agriculture can go some way to filling the gap, but there are still some drawbacks that must be overcome.

The final two chapters of the book take a step back for a broad look at the crisis and what is needed in terms of the way ahead. In Chapter 16, Marcia Ishii-Eiteman argues that the current food crisis is a systemic one requiring structural changes to the current global food system. For suggestions on the way ahead, she looks to two recent UN reports—one on the Right to Food from the UN Human Rights Council, and the other from the UN-led International Assessment of Agricultural Knowledge, Science, and Technology for Development. She argues that there must be a strengthening of the small-scale farm sector, an increase in investment in agroecological methods, more equitable international agricultural trade agreements, and enhanced local participation in agricultural decision making. Finally, in Chapter 17, Alex McCalla identifies four major challenges that must be overcome if we are to achieve global food security: the need to increase productivity, the need to liberalize agricultural trade, the need for national governments to interface their own agricultural systems with world food markets, and the need for better national and international food safety nets. He argues that, while all are important, productivity increases are a priority, warning that we cannot neglect investment in agriculture again, as was the case in recent decades, or we will risk future crises that could be far worse.

The chapters in this book represent a range of viewpoints. As was apparent at the workshop in which the chapters were first presented, there are a number of areas where authors expressed differing opinions. Especially controversial issues included the extent to which financial speculation was a leading driver in the food price spikes, whether agricultural trade should be liberalized or more closely managed by governments, the role of genetically modified organisms in a future agricultural development framework, the implications of a growing role for private corporations in the global food system, and the usefulness of globally managed grain stocks as a means to prevent future food crises.

Despite the differences, there are some areas in which there is agreement among the authors in terms of the way forward. In the immediate term, more financial support is required for the WFP to provide emergency food aid. In the longer term, there is a need for vastly increased public investment in small-farm agriculture, particularly in the world's poorest countries. This investment should promote sustainable agricultural methods that recognize the complex linkages between climate change, agricultural productivity, and hunger. Biofuel policies must also be reformulated in ways that enhance both sustainability and food security.

Above all, a key point of agreement among the authors is that it is essential that food and agriculture remain high on the global policy agenda, as they were during the first half of 2008. Although the superficial causes of the 2008 food price spikes abated somewhat by early 2009, the underlying structural forces contributing to the crisis are still present and have not been adequately addressed at the international level. If high-level attention wanes, as happened following the 1970s food crisis, then the number of hungry people will continue to grow and the promises to make concerted efforts to achieve food security for all, made by global leaders at multiple meetings in the past thirty-five years, will remain unfulfilled for yet another generation.

Works Cited

Blas, Javier (2009). "Number of Chronically Hungry Tops 1 bn." *Financial Times*. 26 March.

Food and Agriculture Organization of the United Nations (2008). "The State of Food Insecurity in the World 2008." Rome. ftp://ftp.fao.org/docrep/fao/011/i0291e/i0291e00.pdf.

World Bank (2007). "Agriculture for Development: World Development Report 2008." Washington, DC. http://siteresources.worldbank.org/INTWDR2008/Resources/WDR_00_book.pdf.

——— (2008). "Rising Food Prices: Policy Options and World Bank Response." Washington, DC. http://siteresources.worldbank.org/NEWS/Resources/risingfoodprices_backgroundnote_apr08.pdf.

World Food Programme (2009). "Weathering the Storm: Coping with High Food Prices and the Financial Crisis." Rome. http://home.wfp.org/stellent/groups/public/documents/newsroom/wfp197596.pdf.

PART 1

The Causal Factors behind the Food Crisis

CHAPTER 2

The Blame Game

Understanding Structural
Causes of the Food Crisis

Anuradha Mittal

While the latest global forecasts show that food prices are finally stabilizing after the spikes of early 2008, the crisis is far from over. Forecasts from the Food and Agriculture Organization of the United Nations (FAO), Organization for Economic Co-operation and Development (OECD), and the US Department of Agriculture (USDA) project that the increase in food prices is not a temporary phenomenon and that prices are likely to remain well above the 2004 levels through 2015 for most food crops (World Bank 2008). The impact of food price increases on the poor and on low-income consumers, resulting in discontent and protests, is the most disconcerting aspect of the situation.

A crisis of this magnitude has thus evoked both national and international responses. However a failure to examine the factors driving the crisis is bound to result in failed prescriptions to deal with the situation. This chapter first provides an analysis and critique of the factors most cited as reasons for the price increase and then outlines deeper, longer-term structural causes behind growing hunger.

Short-Term Causes of the Food Price Crisis

Several factors have been cited regularly as being responsible for the increase in food commodity prices. These include:

Decline in Growth of Agricultural Production

The global supply and demand for food commodities has been affected by both long-term and short-term factors, which have slowed growth in production on one hand and strengthened demand on the other, causing agricultural prices to increase.

Compared to the period between 1970 and 1990, when the production of aggregate grains and oilseeds rose by an average of 2.2 percent per year, the growth rate has declined to about 1.3 percent since 1990. The growth rate of grain production is estimated to decline further to 1.2 percent per year between 2009 and 2017 (World Bank 2008, 5).

Many factors have caused the gradual slowing of production growth. These include a reduction of public support and state intervention in the agricultural sector of the developing countries, a reduced overall investment in agriculture, and a decline in research by governmental and international institutions.

Resource scarcity issues, notably climate change and water depletion, have also affected production growth. Water scarcity is increasingly dire, and each year, 5 to 10 million hectares (25 million acres) of agricultural land are lost because of degradation caused by water shortages (Stigset 2008). Droughts, floods, and freezing weather due to climate change have reduced agricultural output and therefore food security in developing countries.[1]

Figure 2.1

Total World Grain and Oilseeds

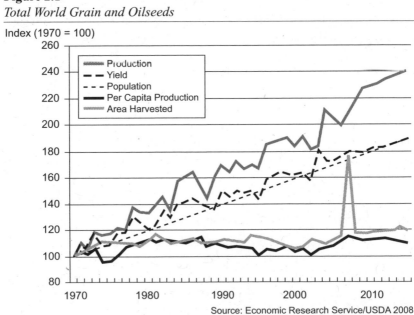

Source: Economic Research Service/USDA 2008

Table 2.1

Exponential Trend Growth Rates

	1970–1990	1990–2007	2009–2017
Production	2.2	1.3	1.2
Yield	2.0	1.1	0.8
Population	1.7	1.4	1.1
Per Capita Production	0.56	0.11	0.2
Area Harvested	0.15	0.14	0.39

Source: Trostle 2008 (USDA data)

Decline in Global Stocks of Grains

The decline in production growth has accompanied a drop in global stocks of grain. According to the USDA, the global grain stocks have declined from 31.2 percent of total global grain production in 1999–2000 to 16.5 percent in 2007–08—the lowest level since 1973.

Several factors are responsible for the low grain stocks. Because the cost of holding grain stocks is as high as 15–20 percent of the value of stock per year (Lin 2008), government-held buffer stocks have come to be considered inefficient and less important after nearly two decades of low and stable prices. Also, some have viewed the liberalization of agricultural markets as reducing the need for individual countries to hold public grain reserves. As the USDA has noted, "for the private sector the cost of holding stocks, use of 'just-in-time' inventory management, and years of readily available global supplies has provided incentives to reduce stock holdings" (Trostle 2008).

In addition, agricultural production is weather-sensitive and a drought or flood can significantly reduce output. For instance, adverse weather conditions impacted some major grain- and oilseed-producing areas in 2006 and 2007. The declining stocks are considered one of the fundamental factors triggering the initial spur of speculative demand in recent years, further fuelling the price hikes (ADB 2008).

These factors are further compounded by the concentration of the international food trade in the hands of a few transnational corporations that, operating within a skewed international trading system, ensures that the majority of profits flow to the global North.

Increasing Energy Costs Spur Production Costs

According to the USDA's cost-of-production surveys and forecasts, production costs for US corn, soybeans, and wheat increased by approximately 21.7 percent between 2002 and 2007, driven by the near doubling of prices of energy intensive components of production, including fertilizer and fuel. Production cost increases in turn led to a 15–20 percent increase in export prices of major US food commodities between 2002 and 2007 (Mitchell 2008).

Increased Demand from the Emerging Economies

The surge in food commodity prices has also been attributed by some, such as the International Monetary Fund (IMF), to "strong per capita income growth in China, India, and other emerging economies," which "buoyed food demand, including for meats and related animal feeds, especially grains, soybeans, and edible oils" (IMF 2008). Then US President George W. Bush referred to a "350 million"-strong middle class in India to argue that its demand for better nutrition was a factor pushing up the global food prices (Prasad and Mittal 2008). The USDA has, in turn, pointed to the "China factor" (Jiang 2008).

It seems highly probable that mass consumption in India and China, which experienced economic growth at a rate of 9.2 percent and 11.4 percent respectively in 2007 and which together account for nearly one third of world's population, could create a food crisis. A closer examination, however, reveals that this is not the case.

Demand for food is income inelastic—the quantity people demand for food does not vary much with changes in incomes, as people need a certain quantity of food to survive, but, at the same time, they cannot consume much more than what is enough for them. What changes is the composition of the food basket. Increased incomes lead to demand for higher quality food, and, in this case, meat. This is explained by Bennett's law, which states that the share of animal products in calories consumed increases as incomes rise (Poleman 1981).

This argument is questionable, however, in the case of India, where the consumption of red meat is low for cultural and religious reasons. There has been extraordinary growth in the consumption of milk, eggs, and poultry meat, but per capita consumption of these products is low: 37 eggs and 1 kg of poultry meat per capita per annum. Also, poultry is one of the fastest growing segments of the agricultural sector in India today with the production of eggs and broilers rising at a rate of 8–10 percent per annum. As a result, India is now the world's fifth largest egg producer and the eighteenth largest producer of broilers (Delgado and Narrod 2002).

In addition, both India and China, while maintaining a food trade surplus, remain net exporters of cereals. China's food trade balance maintained an average surplus of US$4 billion from 2000 to 2006 and has always been a net exporter of cereals (Berthelot 2008b). India, too, has been a net exporter of agricultural and food products since 1995. It is also a net exporter of meat and dairy products. In contrast, the European Union (EU) remains the largest importer of oil seeds, fifth largest importer of cereals in 2007–08, and its food trade balance remains in deficit (Berthelot 2008a).

A report from the World Bank, which attributes rising prices to the production of biofuels, also vindicates the developing countries in their role behind the food price crisis. It states: "Increase in grain consumption in developing countries has been moderate and did not lead to large price increases. Growth in global grain consumption (excluding biofuels) was only 1.7 percent per annum from 2000 to 2007, while yields grew by 1.3 percent and area grew by 0.4 percent, which would have kept global demand and supply roughly in balance" (Mitchell 2008, 17).

Speculation in Financial Markets

The influx of hedge funds, index funds, and sovereign wealth funds in agricultural commodity markets has been suggested as one of the key forces behind the hyperinflation of basic food staples in the short run. It is also one of the two factors that differentiate the current crisis from the previous ones as the nature of the speculating actors has changed significantly.

The futures market is supposed to be a "stabilizing" tool for farmers to sell their harvests ahead of time. In a futures contract, quantities, prices and delivery dates are fixed, sometimes even before crops have been planted. Because speculators are supposed to buy when prices are low and sell when prices are high, they serve to make prices *less* volatile rather than more so.

However, deregulation allowing Wall Street banks to speculate in agricultural futures contracts in essentially unlimited quantities through "swaps" agreements, along with the systematic exploitation of regulatory loopholes, has facilitated a surge in speculative investment in commodity markets to unprecedented levels in recent years.[2] Additionally, speculators are often no longer directly involved in agricultural production or processing but rather are "noncommercial" hedge funds and indexes. Investment in commodity-futures-indexes, primarily from non-commercial speculators, grew from US$13 billion at the end of 2003 to US$260 billion as of March 2008 (Masters 2008). Also, with the burst of the US housing bubble and global grain stocks growing low, financial speculators saw opportunities in the food commodities markets to diversify their financial portfolios and improve returns for their investors.

The artificial demand created by investors' speculation in commodities futures put tremendous upward price pressure on food and energy commodities. For instance, wheat, a commodity that has been increasingly subject to speculative trade in the commodity futures exchanges, along with corn, rice, and soya, has experienced extreme price volatility. Wheat prices have seen a series of spikes and dives since January 2008: between 10 January and 26 February 2008, prices increased by 46 percent, fell an equal amount by May 19, increased again by 21 percent to early June, and then fell until August (Ghosh 2008). The rapidity and volitility of changes suggest forces other than supply and demand at play.

In June, the US Homeland Security and Governmental Affairs Committee held pension funds responsible for price spikes in commodities markets. The huge influx of fund money into commodity markets between 2003 and 2008 paralleled aggregate commodity price increases greater than those of any other period in recorded US history. In a dramatic bid to lower food and energy prices, the committee proposed barring schemes with more than US$500 million in assets from investing in US agricultural and energy commodities. The proposed bill, the Commodity Speculation Reform Act of 2008, passed in the US House of Representatives with major revisions in September 2008 (Institute for Agriculture and Trade Policy 2009).

Biofuels

Perhaps the most prominent difference between the current food price crisis and earlier ones is the increase in demand for grain due to biofuels production in the United States and European Union. Although there is general consensus that demand for biofuels has played a proximate role in price spikes by tightening the global market, the United States has argued for their more marginal role against the position of the World Bank and the OECD (Trostle 2008). Biofuels and the related consequences of low grain stocks, large land-use shifts, speculative activity, and export bans, have been held responsible for a 70–75 percent increase in food prices (ibid.). While Brazil is also a significant producer of biofuels, it is estimated that its sugar-cane-based ethanol production has not contributed appreciably to the increase in food commodity prices (Mitchell 2008).

High oil prices in recent years, together with concerns over energy security and climate change, have promoted the production and use of biofuels as a supplement to transportation fuels. Generous policy support (subsidies and tariffs on imports) and ambitious mandates—the 2007 US Energy Bill almost quintupled the biofuels target to 35 billion gallons by 2022, while the EU wants 10 percent of transportation fuels to be provided by biofuels by 2020—provided a further boost.

The EU, the largest biodiesel producer, began to increase biodiesel production in 2005,[3] while US ethanol production has risen more rapidly since 2002.[4] Between 1980 and 2002, the amount of corn used to produce ethanol in the United States rose by 24 million metric tons, accounting for 7 percent in the total increase in the demand for wheat and coarse grains.[5] However, between 2002 and 2007, the quantity of US corn used to produce ethanol increased by 53 million metric tons, accounting for 30 percent of the global growth in wheat and feed grains use (Trostle 2008, 15).

As ethanol production has expanded, corn stock levels have declined and corn prices have increased. According to Keith Collins (2008), chief economist at the USDA, the US stock-to-use ratio of corn has dropped from a 24 percent average (1980 to 2004) to 11.1 percent in 2007–08, or a little over one month's supply. "In 2008/09, it is expected to drop to 5.4 percent, only 20 days of supply and the second lowest level in 49 years of records." According to Collins, "there is little prospect of a return to the historical ratio because demand for corn is increasing, and the market is tight. Simply stated, the U.S. and global grain economies are at risk."

Without these increases, Mitchell (2008) estimates, "global wheat and maize stocks would not have declined appreciably, oilseed prices would not have tripled, and price increases due to other factors, such as droughts, would have been more moderate. Recent export bans and speculative activities would probably not have occurred because they were largely responses to rising prices."

Land-use changes due to expansion of acreage under biofuels feed-stocks have reduced production of other crops. For instance, US rice production decreased by 12 percent from 2006 to 2007 after 16 percent of the area sown in rice was moved to corn (Berthelot 2008a, 8). Corn expansion also resulted in a 16 percent decline in soybean planting, thereby reducing soybean production and leading to a 75 percent rise in soybean prices between April 2007 and April 2008 (Mitchell 2008).

The EU's expansion of biodiesel production diverted land from wheat to oilseeds, slowing the increase in wheat production. The eight largest wheat-exporting countries expanded area in rapeseed and sunflower by 36 percent between 2001 and 2007, while wheat area fell by 1.0 percent. The wheat production potential of this land was 26 million tons in 2007 and totalled 92 million tons from 2002 to 2007 (Mitchell 2008).

With few countries responsible for exporting staple cereal grains such as corn, rice, and wheat, least developed countries (LDCs) and developing countries have come to largely rely on imports from these countries. So any changes in the policies of major cereal-exporting countries have a significant impact on the world markets. Since the United States is the world's largest corn exporter, higher prices resulting from increased US demand for biofuel production have spilled over onto world markets, triggering an international crisis.

Long-Term Structural Factors behind the Food Price Crisis

Short-term factors outlined above triggered lower supplies and resulted in food price increases. It is essential to understand how developing countries have come to be vulnerable to supply changes caused by short-term factors. However, failure to examine structural causes that have been at work for the last few decades would present an incomplete and incorrect picture of the current crisis and prevent appropriate action to alleviate the problem.

Decline in Investment in Agricultural Productivity

Despite evidence indicating that investment in agriculture results in positive growth and poverty reduction, spending on farming as a share of total public spending in developing countries fell by half between 1980 and 2004 (Jiang 2008). The situation is especially severe in sub-Saharan Africa (Akroyd and Smith 2007). This trend started during the 1980s and 1990s when the World Bank's Structural Adjustment Loans promoted reforms in the agricultural and financial sector. These reforms aimed at reducing the role of the public sector in agricultural marketing, removing agricultural input and food subsidies, withdrawing specialized credit facilities for agriculture, and downsizing agricultural sector agencies—which included eliminating national grain reserves in many instances and closing down marketing boards—as conditions for receiving new loans or restructuring existing debt.

Deregulation of the financial sector, which led to the closure of rural bank branches, created an urban bias in loan allocation and shifted rural savings to urban and commercial credit as the market-oriented financial sector responded to short-run return differentials, adversely impacting financing for agriculture (Chowdhury 2002). The overall impact in most countries was that government expenditure in agriculture fell sharply. Poor public investment, in turn, led to a lack of private investment (Cleaver and Donovan 1995). In several countries, failure to adhere to IMF and World Bank conditionalities triggered temporary (and sometimes permanent) postponements of cash releases and changes in commitments from other donors that further destabilized the level of expenditure in the agricultural sector (ibid.).

These externally imposed mandates prevented developing countries, especially those in sub-Saharan Africa, from making needed investments in agriculture. National government funding of agricultural science fell by 27 percent in sub-Saharan Africa between 1981 and 2000, with many governments currently allocating less than 1 percent of their national budgets to the sector (Hanson 2008). In July 2003, members of the African Union agreed to devote at least 10 percent of their government budgets to agriculture programs over the next five years. So far only Rwanda and Zambia have executed the plan.

Countries have reduced and even eliminated support for farm credit, crop distribution, and reserve programs. The elimination of seed and fertilizer subsidies, a keystone of World Bank austerity policies, resulted in African farmers abandoning higher-yield seeds leading to a decline in crop yields and production. When Zambia eliminated corn seed and fertilizer programs, corn acreage and fertilizer application both declined sharply (World Bank 2002).

At the same time, multilateral investment in agricultural projects in poor countries and agricultural research by the governments of rich nations and institutions such as the World Bank have also been declining (Jomo 2008). Agricultural research grants were cut in half—from US$6 billion to US$2.8 billion annually—between 1980 and 2006, with the United States alone decreasing its contribution from US$2.3 billion to US$624 million. USAID, the US international development agency, has cut its agricultural aid by 75 percent in the past two decades; just 4 percent of current development aid to Africa goes to investment in agriculture. The World Bank decreased its lending for agriculture from US$7.7 billion in 1980 to US$2 billion in 2004 (ibid.). The Independent Evaluation Group report on the Bank's agricultural programs in sub-Saharan Africa between 1991 and 2006 states that the Bank channelled US$2.8 billion to agriculture, constituting just 8 percent of its investment lending to the region (World Bank Independent Evaluation Group 2007).

This underinvestment in agriculture by national governments and international donors, and their imposed conditionalities, has prevented adequate farm programs in the poorest developing countries. This has eroded their ability to maintain agricultural production and increased their reliance on imported food.

Reduced State Regulatory Role in Agricultural Production and Trade

During the 1970s, especially in Africa, the World Bank promoted the development and support of a variety of agricultural marketing and processing parastatals. In the 1980s and 1990s, it strongly encouraged the withdrawal of the state's regulatory role, for instance through the elimination of agricultural marketing boards.

Marketing boards were made responsible for managing the stock of food at the national level. They bought agricultural commodities from farmers at a price fixed high enough to cover the cost of production plus a profit, keep the commodities in a rolling stock, and release them into the market in the event of a bad harvest in following years. Marketing boards also organized the redistribution of food from surplus to deficit areas of the country. Preventing price volatility, marketing boards protected both producers and consumers against sharp rises or drops in prices, prioritized self-sufficiency, and therefore reduced the need for food imports and for foreign currency.

However, marketing boards had their problems. In many developing countries, especially in Africa, they were found inefficient, over-staffed, and often corrupt. Inefficiencies in the state-run marketing system squeezed farm-gate prices and burdened state budgets. Thus, the donor/lender-sponsored reform or elimination of marketing boards appeared reasonable, especially from the point of view of balancing the state budget.

After over two decades of economic liberalization and related reforms, however, the promised or expected gains in terms of growth and stability have not been realized. The recent food crisis and the vulnerability of food security in developing countries point to the fact that the goals of state intervention, particularly in staple crop marketing, remain valid. Therefore, the reform agenda should have aimed at improving the efficiency and reducing the waste associated with the marketing boards instead of closing them down.

Removal of Agricultural Tariffs and Resulting Import Surges

A recent fact sheet from the US Trade Representative's (USTR) office states: "Trade is a powerful tool to generate income gains that can dwarf foreign assistance. …The World Bank estimates that low and middle income countries would realize 50 percent of their potential economic gains from global free trade in goods, by the elimination of their own barriers" (2008). However, the indiscriminate opening of markets has taken away the ability of developing countries to govern the flow of agricultural imports into their markets.

Heavily subsidized agriculture has allowed industrialized countries to capture developing country markets by dumping commodities below the cost of production.[6] In 2003, the United States exported wheat at 28 percent below the cost of production, soybeans at 10 percent below the cost of production, corn at an average price of 10 percent below the cost of production, cotton at 47 percent below the cost of production and rice at 26 percent below the cost of production (Murphy, Lilliston, and Lake 2005).

The flood of cheap farm imports, often from countries where agriculture is heavily subsidized, has made subsistence farming production in many developing countries—especially in Africa—uncompetitive and financially unsustainable, resulting in farmers leaving or being forced off the land. This process of "deagrarianization"[7] has turned some of these countries from net exporters to large importers of food, directly threatening their food security and economic sustainability.

The FAO Briefs on Import Surges document up to 12,167 import surges between 1980 and 2003 in 102 developing countries, with "devastating consequences for the rural poor and local economies in Africa" (Kwa 2008). Food import surges have affected developing countries everywhere, including South

Box 2.1 *The Experience of Ghana*

From the 1960s to the 1980s, Ghana's policies to promote self-sufficiency in food involved the government actively encouraging the agricultural sector through marketing, credit, and subsidies for inputs.

But under pressure from the World Bank and IMF—from the mid-1980s onward and especially in the 1990s—the policies for self-sufficiency were reversed. Input subsidies were eliminated, the state trading enterprise (Ghana Food Distribution Corporation) was phased out, and the system of minimum guaranteed prices for rice and wheat was abolished, as were many state agricultural trading enterprises and the seed agency responsible for producing and distributing seeds to farmers. Loans from the commercial banks to the agricultural sector dwindled from 13.6 percent in 1993 to 1 percent by 2004. At the same time, applied tariffs for most agricultural imports were reduced significantly to the present 20 percent. These measures left local farmers unable to compete with imports artificially cheapened by high subsidies, especially in rice, tomato, and poultry.

Ghana's rice imports increased from 250,000 metric tons in 1998 to 415,150 metric tons in 2003, an increase of 70 percent. Domestic rice, which had accounted for 43 percent of the domestic market in 2000, captured only 29 percent of the domestic market in 2003. In all, 66 percent of rice producers recorded negative returns, leading to loss of employment. Rice farmers were squeezed out of the market along with other players in the value chain—traders, millers, transporters, and so on. In response, the government raised tariffs on rice imports from 20 percent to 25 percent, but the increase was in place for only four days before it was removed under pressure from the IMF. In the same year, 2003, the US government provided domestic rice subsidies worth US$1.3 billion. A government study found that 57 percent of US rice farms would not have covered their cost had they not received subsidies. In the period 2000–03, the average cost of production and milling of US white rice was US$415 per metric ton, but it was exported for just US$274 per metric ton, a price 34 percent below cost.

Ghana's poultry sector was at its prime in the late 1980s but declined steeply in the 1990s due to the withdrawal of government support and the reduction of tariffs. Poultry imports rose by 144 percent between 1993 and 2003, and heavily subsidized poultry from Europe made up a significant share of these imports... Between 1996 and 2002, EU frozen chicken exports to West Africa rose eight-fold, due mainly to import liberalization, practically wiping out the half million chicken farmers in Ghana. In 1992, domestic farmers supplied 95 percent of Ghana's market, but this share fell to 11 percent in 2001. In 2003, Ghana's parliament raised the poultry tariff from 20 to 40 percent. This was still much below the bound rate allowed by the World Trade Organization (WTO) of 99 percent. However, the IMF objected to this move and the new approved tariff was not implemented (see Khor 2008).

and Southeast Asia, Latin America, and the Caribbean. Although each country is affected in different food markets, the narratives are strikingly similar: an import surge of a food staple displaces the domestic market, thereby decreasing domestic production and employment by startling percentages.

According to the United Nations Conference on Trade and Development (UNCTAD, 2008), current high international food prices are expected to bring about another episode of food import surges, which have become more frequent in the LDCs in the post-trade liberalization era. Countries whose local agricultural base was impacted by the dumping of cheap grains, in the form of food aid and cheap subsidized commodities from richer nations, are now experiencing shortages because the markets they have come to depend on have changed their policies. The US and European biofuel policy is a case in point: corn production dedicated to biofuels instead of food compounds scarcity in both the market availability and food aid availability of the grain.

Shift to Export Crops

An estimated forty-three developing countries, of which three-quarters are LDCs, depend on a single commodity (sugar, coffee, cotton lint, or bananas) for more than 20 percent of their total revenues from merchandise exports (FAO 2004). Governments in these countries have failed to restructure their economies, which still have the legacies of colonial plantation-based production and trade structures. The policy advice of donors/lenders has reinforced this structure claiming comparative advantage.

However, the real prices of these commodities are volatile, and, as a direct consequence, these countries are subject to significant risk, which affects the level of macroeconomic activity as well as the households' income distribution (Bourguignon, Lambert, and Suwa-Eisenmann 2004). For example, coffee prices fell in 2002 to less than a third of their 1997 level. Uganda, a country that implemented the trade and economic reforms requested of it in the 1990s and increased coffee production, saw many of the gains undermined, if not wiped about, by a decline of world coffee prices that were beyond its control.

According to the FAO, "declines and fluctuations in export earnings have battered income, investment and employment in these countries and left many of them deeply in debt." Thirty-seven out of the forty-two countries identified as Heavily Indebted Poor Countries (HIPCs) by the IMF and World Bank rely on primary commodities for more than half of their merchandise export earnings. More than half the world's cocoa and more than a quarter of its coffee are produced in countries classified as HIPCs (FAO 2004). The FAO also contends that if prices for the ten most important (in terms of export values) agricultural commodities had risen in line with inflation since 1980, these exporters would have received around US$ 112 billion more in 2002 than they actually did. This is more than twice the total amount of aid distributed worldwide (ibid.).

This specialization in a few commodities such as coffee or cocoa has created an increased dependence on food imports from developed countries and converted developing countries from net food exporters to net food importers (ibid., 14). "In the 1960s, developing countries had an overall agricultural surplus of US$7 billion. By the 1970s, imports had increased and the surplus had shrunk to US$1 billion. By the end of the 1980s, however, the surplus had disappeared. Most of the 1990s and 2000s saw developing countries develop into net food importers. The deficit in 2001 was US$11 billion" (ActionAid 2008).

Africa has been particularly impacted by the liberalization of markets and diversion of resources from food crop production to cash crop investments, adding twice as many acres of new cotton production as new acres of corn and fifty percent more new acres of cocoa beans than new acres of millet since the World Trade Organization (WTO) was formed in 1995 (Food & Water Watch 2008). In the absence of international markets for traditional African crops like sorghum, cassava, yams, and millet, farmers have been encouraged to grow cash crops like coffee, sugar, cocoa beans, tea, and cotton. Export earnings are used to purchase food, often low-priced imports from industrialized countries even as this practice displaces small farmers. With prices of imported food rising, there is insufficient domestic production to provide food for local markets in many countries.

Conclusion

Chronic hunger afflicts hundreds of millions of people. Between 2003 and 2005, the FAO estimated that 848 million people were undernourished worldwide (FAO 2008b). It also reported that the number has been increasing at the rate of almost 4 million per year since the second half of the 1990s, rendering the 1996 World Food Summit goal to halve the number of undernourished people by 2015 a far-fetched idea.

This already grave situation of global hunger was worsened by the 83 percent increase in global food prices between 2005 and 2008. According to the FAO, an additional 75 million people have been plunged below the hunger threshold, bringing the estimated number of undernourished people worldwide to 923 million in 2007 (ibid.).

The crisis of global hunger today needs to be a wake-up call for nations to recognize that agriculture is fundamental to the well-being of all people, both in terms of access to safe and nutritious food and as the foundation of healthy communities, cultures, and environment.

Governance reforms should include restructuring the WTO to create policy space for poor countries to ensure domestic food security, while reducing agricultural subsidies in the rich world. Conclusion of the Doha Development Round of negotiations will only offer a way forward from the food crisis if the embedded inequalities in the current system are addressed.

Urgent action is necessary and will require both a short- and long-term approach. A comprehensive understanding of the causes of the food price crisis is crucial, however, before steps can be taken to ensure global food sovereignty.

Notes

1 For instance, FAO reports that multiple-year droughts caused "exceptional shortfall in aggregate food production/supplies" in Lesotho and Swaziland. In Nigeria and Ghana, the decline of coarse grain production led to tight food supply that affected rising food prices in Benin, Burkina Faso, Ghana, Niger, Nigeria, and Togo. In China's harshest ice rains, snow, and freezing weather since 1951, millions of hectares of vegetable and oil crops were "severely damaged," and "as of the end of January [2008], about 90 million people were reported to be directly affected." In Mongolia, the harsh winter impacted livestock production as well. The villages of the Northern Atlantic Autonomous Region in Nicaragua, affected by powerful hurricane Felix in September 2007, are receiving international food assistance for the gradual recovery of their livelihood systems (FAO 2008a, 2–4).

2 In 2000, the Commodity Futures Modernization Act effectively deregulated commodity trading in the United States by exempting over-the-counter (OTC) commodity trading (outside of regulated exchanges) from CFTC oversight. Soon after this, several unregulated commodity exchanges opened. These allowed any and all investors, including hedge funds, pension funds, and investment banks, to trade commodity futures contracts without any position limits, disclosure requirements, or regulatory oversight.

3 United States uses nearly all corn as a feedstock, while the EU, the largest biodiesel producer, uses rapeseed oil is its main feedstock.

4 Ethanol production jumped from 1 billion gallons in 2005 to 5 billion in 2006 and will reach 9 billion in 2009.

5 Ethanol is produced from sugar crops, such as sugar cane or beets, or starchy crops such as maize. Biodiesel is produced from vegetable oils or animal fats.

6 In fiscal year 2008, US agricultural exports are expected to reach a record US$108.5 billion—US$26.6 billion above 2007.

7 The inability to make a living in agriculture is driving more and more people in LDCs to seek work in other sectors of the economy, a process described as "deagrarianization" (see UNCTAD 2008).

Works Cited

ActionAid (2008). "Impact of Agro-import Surges in Developing Countries." London. http://www.actionaid.org/main.aspx?PageID=202.

Akroyd, Stephen, and Lawrence Smith (2007). "Review of Public Spending on Agriculture." Washington, DC: DFID/World Bank. http://www1.worldbank.org/public sector/pe/pfma07/OPMReview.pdf.

Asian Development Bank (2008). "Soaring Food Prices: Response to the Crisis." Manila. http://www.adb.org/Documents/Papers/soaring-food-prices/soaring-food-prices.pdf.

Berthelot, Jacques (2008a). "Sorting the Truth out from the Lies in the Explosion of World Agriculture Prices." Toulouse: Solidarité.

——— (2008b). "The Food Prices Explosion: False and Actual Culprits." Presented at Assembly of the Enlarged Council of the World Forum for Alternatives. Caracas: World Forum for Alternatives. 13–18 October.

Bourguignon, François, Sylvie Lambert, and Akiko Suwa-Eisenmann (2004). "Trade Exposure and Income Volatility in Cash Crop Exporting Developing Countries." *European Review of Agricultural Economics* 31, no. 3: 369–87.

Chowdhury, Anis (2002). "Politics, Society, and Financial Sector Reform in Bangladesh." *International Journal of Social Economics* 29, no. 12: 963–88.

Cleaver, Kevin M., and W. Graeme Donovan (1995). "Agriculture, Poverty, and Policy Reform in Sub-Saharan Africa." World Bank Discussion Papers, no. 280. Washington, DC: World Bank. http://www-wds.worldbank.org/servlet/WDSContentServer/WDSP/IB/1995/02/01/000009265_3970311122520/Rendered/PDF/multi_page.pdf.

Collins, Keith (2008). "The Role of Bio-fuels and Other Factors in Increasing Farm and Food Prices: A Review of Recent Developments with a Focus on Feed Grain Markets and Market Prospects." Supporting material for a review conducted by Kraft Foods Global, Inc., 19 June. http://www.foodbeforefuel.org/facts/studies/role-biofuels-and-other-factors-increasing-farm-and-food-prices.

Delgado, Christopher, and Clare Narrod (2002). "Impact of Changing Market Forces and Policies on Structural Change in the Livestock Industries of Selected Fast-Growing Developing Countries." Rome: Food and Agriculture Organization.

Food & Water Watch (2008). "What's Behind the Global Food Crisis? How Trade Policy Undermined Africa's Food Self-Sufficiency." Washington, D.C. http://www.foodandwaterwatch.org/food/whats-behind-the-global-food-crisis.

Food and Agriculture Organization of the United Nations (2004). "The State of Agricultural Commodity Markets 2004." Rome. ftp://ftp.fao.org/docrep/fao/007/y5419e/y5419e00.pdf.

——— (2008a). "Crop Prospects and Food Situation: Countries In Crisis Requiring External Assistance" Rome. http://www.fao.org/docrep/010/ah881e/ah881e02.htm.

——— (2008b). "Hunger on the Rise: Soaring Prices Add 75 Million People to Global Hunger Rolls." Rome. 30 September.

Ghosh, Jayati (2008). "The Commodity Price Roller Coaster." New Delhi: International Development Economic Associates.

Hanson, Stephanie (2008). "Backgrounder: African Agriculture." *The New York Times.* 28 May.

Institute for Agriculture and Trade Policy (2009). "Betting Against Food Security: Futures Market Speculation." Minneapolis. http://www.iatp.org/tradeobservatory/library.cfm?refID=105065.

International Monetary Fund (2008). "World Economic Outlook: Housing and the Business Cycle." Washington, DC. http://www.imf.org/external/pubs/ft/weo/2008/01/pdf/text.pdf.

Jiang, Hui (2008). "Rising Agricultural Commodity Prices: How We Got Here and Where Do We Go." Washington, DC: United States Department of Agriculture/ Foreign Agricultural Services.

Jomo, Kwame Sundaram (2008). "Washington Rediscovers Agriculture: The Political Economy of the Agrarian Turn." Keynote address, presented at Peter Wall Summer Institute Public Gala Event, University of British Columbia, Vancouver. 23 June.

Khor, Martin (2008). "The Impact of Trade Liberalization on Agriculture in Developing Countries: The Experience Of Ghana." Malaysia: Third World Network.

Kwa, Aileen (2008). "Why Food Import Surges Are an Issue at the WTO." *IPS*. 7 March.

Lin, Justin (2008). "Preparing for the Next Global Food Price Crisis." Remarks delivered at the Roundtable on Preparing for the Next Global Food Price Crisis, Center for Global Development, Washington, DC. 6 October.

Masters, Michael (2008). Testimony before U.S. Senate Committee on Homeland Security and Governmental Affairs. 20 May.

Mitchell, Donald (2008). "A Note on Rising Food Prices." Policy Research Working Paper no. WPS 4682. Washington, DC: World Bank.

Murphy, Sophia, Ben Lilliston, and Mary Beth Lake (2005). "WTO Agreement on Agriculture: A Decade of Dumping. United States Dumping on Agricultural Markets." Minneapolis: Institute for Agriculture and Trade Policy. http://www.un-ngls.org/ orf/cso/cso7/library.pdf.

Office of the United States Trade Representative (2008). "Trade Facts: The Benefits of Trade for Developing Countries." Washington, DC. July. http://www.ustr.gov/ assets/Document_Library/Fact_Sheets/2008/asset_upload_file226_15014.pdf.

Poleman, Thomas T. (1981). "Quantifying the Nutrition Situation in Developing Countries." *Food Research Institute Studies* 18, no. 1: 1–58.

Prasad, Indulata, and Anuradha Mittal (2008). "The Blame Game: Who Is Behind the World Food Price Crisis." Oakland, CA: Oakland Institute.

Stigset, Marianne (2008). "Food Price Gains Caused 50 Million More to Go Hungry, FAO Says," Bloomberg.com. 3 July.

Trostle, Ronald. (2008). "Global Agricultural Supply and Demand: Factors Contributing to the Recent Increase in Food Commodity Prices." Washington, DC: United States Department of Agriculture.

United Nations Conference on Trade and Development (2008). "Least Developed Countries Report 2008." Geneva. http://www.unctad.org/en/docs/ldc2008_en.pdf.

World Bank (2002). "Implementation Completion Report on a Credit in the Amount of SDR 41.2 to the Republic of Zambia for an Agricultural Sector Investment Program." Report no. 24444. Washington, DC. 30 June.

———— (2008). "Rising Food Prices: Policy Options and World Bank Response." Washington, DC. http://siteresources.worldbank.org/NEWS/Resources/risingfoodprices _backgroundnote_apr08.pdf.

World Bank Independent Evaluation Group (2007). "Assistance to Agriculture in Sub-Saharan Africa: An IEG Review." Washington, DC. http://siteresources. worldbank.org/EXTASSAGRSUBSAHAFR/Resources/ag_africa_eval.pdf.

CHAPTER 3

The 1974 and 2008 Food Price Crises
Dèjà Vu?

Sue Horton

S harply rising grain prices, food riots, increasing oil prices, low food stocks, low food aid deliveries, food export restrictions/bans, and blame assigned to speculators—are we talking about 2008 or 1974? The similarities between the two food price crises are striking. Although the world changed considerably between 1974 and now, as far as international food markets are concerned, many of the same problems persist.

Did we not learn anything from the 1974 food crisis? Why did the solutions implemented in 1974 not prevent a recurrence? What can we do to prevent another future food price crisis?

This paper first discusses the similarities (and differences) between the 1974 food price crisis and the current one. I then provide my personal assessment of the policy options post-2008, taking into account the successes and failures of policies implemented in 1974.

Similarities—and Differences—between the 1974 and 2008 Food Price Crises

The following excerpts from a *Time Magazine* article for 11 November 1974 show strikingly how history repeats itself:

> The world's reserves of grain have reached a 22-year low….Low harvests and high prices have forced the traditional surplus-producing nations to curtail the amount of food that they normally give as aid…Argentina, Brazil, Thailand, Burma and the Common Market nations have restricted food exports.

Against this gloomy backdrop, about 1,000 delegates from some 100 nations and a dozen international organizations are gathering in Rome this week for the World Food Conference.

Food riots have become commonplace in vast sections of Bangladesh and India.

Time also considered the causes of the crisis:

Then came 1972. Bad weather started to plague so much of the world's crop land that many experts conclude the climate itself is changing…Harsh winters, droughts or typhoons cut output in the Soviet Union, Argentina, Australia, the Philippines and India. The weather improved in 1973, but a new set of problems threatened food output…Fertilizer was in short supply, and its price started to climb. Then came the devastating impact of the quadrupling of the market price of petroleum.

With minor changes to the dates, and some—but not all—the country names, this same article could have been reused in 2007–08.

Figure 3.1 plots wheat prices for 1973–74, and 2007–08 on the same graph (January 1973 is the base year for the 1973–74 plot, and January 2007 for the

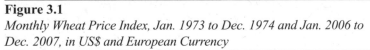

Figure 3.1

Monthly Wheat Price Index, Jan. 1973 to Dec. 1974 and Jan. 2006 to Dec. 2007, in US$ and European Currency

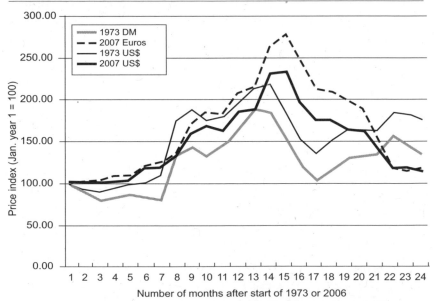

Source: Wheat data for No 1 hard red winter (ordinary protein) wheat, Kansas City MO; USDA-ERS (2009); US $/Euro exchange rate from Federal Reserve Bank (2009); US $ rate from Federal Reserve Bank of St. Louis (2009)

2007–08 plot). Figure 3.2 provides a similar plot for rice. There are some differences in the circumstances of the two crises. In 1973–74, the US dollar was falling against the European currencies, whereas it was rising in the first half of 2007. Therefore the 1973 crisis looks worse using US dollars, whereas the 2007 crisis looks worse using euros. The experience for individual developing countries depends on the major currency with which their own currency is more aligned. However, the similarities between the two crises, in terms of the evolution of prices, are very clear.

Grain Stocks

An analysis of world grain stocks suggests that there was no great surprise that food prices started to rise in 2007, and, given a small supply shock (drought in Australia), that a price crisis ensued. Figure 3.3 shows world wheat stocks at end-year for the period 1974–75 to 2008–09, and also shows that the stock-to-use (consumption) ratio globally reached its lowest level in 2007–08 since 1974–75.

It is expensive to hold food stocks, because food is bulky and perishable. World trade markets in food are thin, relative to many other commodities, and

---

Figure 3.2

Monthly Rice Price Index, Jan. 1973 to Dec. 1974 and Jan. 2006 to Dec. 2007, in US$ and European Currency

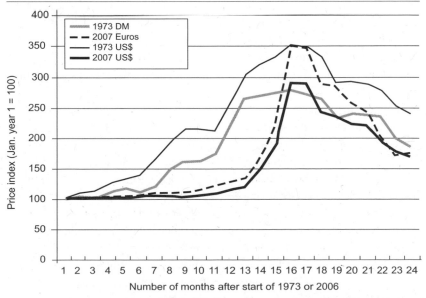

Source: Rice price data Monthly export price (US$/t fob) Thai rice 5% brokens, IRRI 2009. See Figure 1 for sources for exchange rates.

Figure 3.3

End-Year Wheat Stocks, and Stock-to-Consumption Ratio, Wheat (Worldwide), 1970–71 to 2007–08

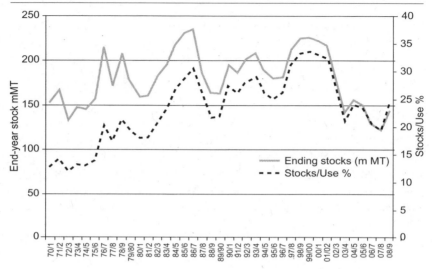

Source: Calculated from USDA (2008)

this is particularly true for rice, a market in which there are few net exporters.[1] Consumer demand for grain in poor countries is inelastic with respect to price, since poor people cannot substitute away from basic staples. Although price increases can send a signal to farmers to produce more staples, this takes a few months to have an effect.

For all these reasons, traded food markets are subject to price volatility, and the low stock-to-use ratio in 1973–74 and again in 2007–08 presented a vulnerability. All it required was an adverse event, or set of events, to trigger a crisis. In the 1970s, the sequence of events started with bad weather in the USSR, Asia, and Africa in 1972, causing world production to drop nearly 40 million metric tons (Hathaway 1975). The USSR then chose to import heavily to offset domestic shortfalls, an unusual practice for them. The United States had been in the process of drawing down their stocks because of storage expense, and due to poor surveillance was caught unaware of the large USSR purchases. Although production recovered in 1973, it was not enough to rebuild stocks. All it took was another year of adverse weather (as was the case in 1974 for both the USSR and the United States), to precipitate a crisis (ibid.).

In 2007, grain stocks were again at record low levels, for reasons to be discussed further below. As in 1974, a major exporter (in 2008, the European Union) had deliberately run down stocks, a consequence of rationalization in the Common Agricultural Policy. The butter and beef mountains (and milk lakes), which had built up due to farm support prices, had dwindled as the Common Agricultural Policy was reformed. All it then took was adverse weather in some major producing areas to push food prices (which had been trending upward in 2005 and 2006) even higher.[2]

Energy Prices

In both 1974 and 2008, oil price rises were a trigger for food price increases. Oil prices rose almost 450 percent from October 1973 to May 1974, partly related to the response of the Organization for Petroleum Exporting Countries (OPEC) to the Yom Kippur war in October 1973, and compounded by a commodities boom (Rogers 2008). Fertilizer prices track oil prices, and in 1974 this was exacerbated by Morocco's decision to treble the price of rock phosphate in the first half of 1974 (ibid.). Although the increase was not of the same magnitude as that of the 1970s, oil prices almost doubled between August 2007 and August 2008, again as part of a commodities boom. Between January 2000 and September 2007, oil and wheat prices tripled, and corn and rice prices doubled (von Braun 2007). The link is not surprising: food production requires energy (for machinery, petrochemical inputs such as fertilizer, and for transporting output).

Increased Demand

These trigger events, however, do not precipitate a food crisis without an underlying imbalance in supply and demand. In 1974, the strong increase in demand resulted from a combination of relatively fast population growth in the developing world and rising demand for meat in the industrialized world. In the developing world, the demographic transition was in its early stages, in which mortality rates were dropping but birth rates had not yet responded (ibid.). The Club of Rome had recently published (in 1972) *The Limits to Growth*, its first report reflecting concerns about population growth and pressure on resources. In industrialized countries, demand for meat was increasing, as detailed in Francis Moore Lappé's *Diet for a Small Planet*. The book highlighted the inefficiency of animal-based diets, reporting that it takes three calories of grain to produce one calorie of meat for human consumption using chickens and pigs, and eight calories using cows (Lappé 1971). The Soviet Union chose to import grain to maintain grain and animal product consumption despite domestic crop downturns.

Although circumstances were different in 2008, the phenomenon of increasing demand was the same. The developing countries were much further along in the demographic transition and birthrates had fallen (although population growth will continue for some decades as a result of the very young population age structure). However, a major driver was rapid economic growth particularly in China and India, where almost one-third of the world's population lives. In these countries, more affluent consumers demanded more animal products—more meat in China, and more dairy (and somewhat more meat) in India; a similar phenomenon was underway in other large developing countries such as Brazil.

One new factor in 2008 was the demand for food grains to produce biofuel. Rising fuel prices—and dwindling oil stocks—led industrialized countries to subsidize the production of biofuel, whose production has risen dramatically since 2000 (IFIF/FAO 2006; Renewable Fuels Association 2008). The United States, for example, has mandates for ethanol blending and tariffs on imported ethanol, as well as subsidies (Rosegrant 2008). The percentage of the US corn crop used for ethanol production rose from 6 percent in the 2001–02 crop year to 18 percent in 2007–08 and 24 percent in 2008–09 (see IFIF/FAO 2006; Kojima and Klytchnykova 2008). Similarly, Brazil has legislation requiring ethanol–gasoline blends and devotes half its sugarcane to ethanol (Kojima and Klytchnykova 2008). The biofuel phenomenon is therefore closely linked to energy prices discussed in the previous section.

The International Food Policy Research Institute's (IFPRI's) IMPACT model suggests that this has had a noticeable effect on food prices. If biofuel demand were frozen at 2007 levels, the model predicts significant drops in food prices (notably maize, 6 percent lower by 2010 and 14 percent lower by 2015), with smaller effects for wheat, cassava, sugar, and oils (Table 3.1, using Rosegrant 2008). If there were a complete moratorium on using agricultural products to produce biofuel, the effect would be even stronger. Maize prices would be 20 percent lower in 2010 and 21 percent lower in 2015, with corresponding double digit drops in 2015 for wheat, sugar, and cassava.

In the long run, it is predicted that second generation biofuels will compete less directly with food consumption as production using sugar cane waste, cellulose, and even algae become economical (ibid.). In the short run, however, US policies are both inefficient (favouring domestic ethanol production using maize, which is more costly than production using sugarcane from Brazil), and compete more directly with food (land used for sugarcane is not as directly substitutable to grain production and world sugar prices remain depressed).

The enthusiasm over biofuel from the earlier 2000s diminished somewhat in 2008 with the sobering realization that—absent additional investments in agriculture—"green" fuel for cars competes directly with feeding poor people. Without putting the brakes on biofuel completely, devising a solution requires renewed attention to agricultural research and careful consideration of distortions (favouring maize use over sugar, for example) and of the overall greenhouse gas effect of biofuels.

Table 3.1

Percentage Change in Selected Crop Prices if Biofuel Demand Were Limited

Crop	Biofuel Freeze 2007 levels Prices in 2010	Biofuel Moratorium Prices in 2010	Biofuel Freeze 2007 levels Prices in 2015	Biofuel Moratorium Prices in 2015
Maize	–6	–14	–20	–21
Wheat	–2	–4	–8	–11
Sugar	–1	–4	–11	–12
Oils	–2	–6	–1	0
Cassava	–2	–5	–14	–19

Source: Trostle 2008 (USDA data)

Speculation and Middlemen

In 1974 and also in 2008, speculators and middlemen were blamed for rising prices. Economists regard trading in grain futures as a generally useful phenomenon, allowing producers to hedge against risks, for example. Obvious abuses such as "cornering" markets (secretly amassing a large enough position to dominate a market in a specific crop) are considered undesirable, but such abuses can be limited by requirements for disclosure and transparency. Clearly, however, speculators can exacerbate short-term price volatility.

Leading up to the 1974 crisis, much blame was attached to sales of subsidized US grain to the USSR described in the press as "the great grain robbery" (Luttrell 1973). At the time, the United States was trying to decrease large and costly grain stocks, and the US Department of Agriculture (USDA) had been managing an export subsidy program. However, as the General Accounting Office concluded in hindsight, there were some problems with the program. Exporters had the option of determining the date when they registered for the subsidy, which made it possible for the USSR to conclude a number of deals before the USDA realized the scale of their purchases. The General Accounting Office likewise argued that "[t]he trading rules and procedures of the USDA were not adequate for dealing with the bargaining power of a foreign state trading monopoly" (ibid., 3). Although this did not lead immediately to a food price crisis, it meant that stocks were unusually depleted leading into the 1974 crisis. Some of the blame also was attached to the large, private, and secretive grain companies such as Cargill: "The company became a prime target when the U.S. government went after the big grain exporters for allegedly manipulating the market. It emerged largely unscathed" (Weinberg and Copple 2002).

In the 2007 crisis, critics pointed the finger at speculators. IFPRI quotes David King, the secretary-general of the International Federation of Agricultural

Producers as saying: "Even if it is difficult to gauge the real impact of this financial speculation, it has certainly played a role in influencing trading prices. Take for example the fact that in a normal year, trading and movements on the wheat futures market in Chicago represent the equivalent of 20 times the annual U.S. wheat harvest. In 2007/2008, these movements represented the equivalent of more than 80 harvests" (IFPRI 2008, 9).

This is, however, a symptom of the problems in grain markets, rather than an underlying cause. Arguably it is more important to protect the "entitlement" of poor consumers to buy food (through the protection of their ability to get work or to use social safety nets) than to spend resources tinkering with futures and hedges. The response of governments to price hikes, such as imposing export bans, is equally damaging to market functioning. In 2008, twenty-nine countries imposed such bans during the crisis, behaviour that also occurred in 1974.

Food Aid

Figure 3.3 displays another lamentable similarity between the 1974 and 2008 crises. Food aid varies almost exactly inversely with food prices (donors tend to set budgets in dollars, such that much less food is provided at times of higher prices). Thus, cereal aid was the lowest in 1973–74 of any year (other than 1988) during the twenty-year period from 1970 to 1990, and similarly food aid was the lowest in 2007 of any year during the seventeen-year period from 1990 to 2007 (having declined fairly consistently in volume since 1999). Thus, far from being a stabilizing force during food crises, food (or cereal) aid dries up exactly when it is most needed.[3] Figure 3.4 also plots the wheat price for the same years and shows the clear inverse relation with food (or cereal) aid.

It is too soon to know what the consequences of the 2008 food crisis will be, and how these will parallel those of the 1974 crisis. We do know that food crises restrict both the quantity and quality of diets of the poor. Isenman (1980) shows using time series data that the elasticity of the death rate with respect to the rice price in Sri Lanka was 0.15. He estimates that the increase in rice prices during the 1974 crisis was associated with an increase in the death rate from 7.7 to 8.9 per thousand. One would expect that many of these deaths were among children, and that the corresponding increase in the infant mortality rate would be considerably higher. There are some predictions of the effect of poorer diet quality in 2008. Bouis (2008) estimates that a 50 percent increase in food prices in Bangladesh will cause a 25 percent increase in anemia rates as households purchase fewer animal products and vegetables in order to maintain staple consumption.

Figure 3.4
World Cereal/Food Aid, Compared to Wheat Price, 1970–2007

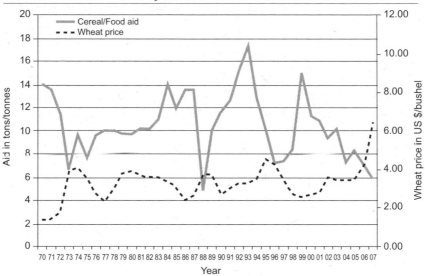

Cereal aid 1970–1990 in million tonnes (July/June year); food aid 1990–2007 in million tons (calendar year) (non-cereal accounted for 1.6m tons in 1990); wheat price is US no. 1 hard red winter (ordinary protein) Kansas City in US $/bushel

Source: Wheat data for No 1 hard red winter (ordinary protein) wheat, Kansas City MO; USDA-ERS (2009); World Food Programme (cereal food aid data)

International Governance, the 1974 Food Crisis, and Possible Responses to the 2008 Crisis

The 1974 food price crisis was a harbinger of further instability in the world economic system; the second oil price increase in 1978 was followed by the debt crisis and a long decade of little growth in the developing world. It is too soon to know what will follow the 2008 food price crisis, but all the current indications are of similar recession and growth slowdown, which will compound the adverse effects on the developing world.

In response to the 1974 food crisis, a World Food Conference was convened in November of that same year. Hathaway (1975) provides a useful summary of the outcomes. The conference emphasized three solutions. The first was to increase food production in developing countries, applying more agricultural inputs and improving policies in order to encourage agricultural production (although Hathaway dryly notes that "the nature of such policies was never spelled out" [70]). The second solution, the responsibility of the Food and

Agriculture Organization of the United Nations (FAO), was a better food security system worldwide, with better information, a system of stocks, and increased food aid. Finally, trade was considered important. Hathaway again comments: "Little was expected on this issue, and the developed countries guaranteed this outcome by insisting trade issues could be discussed only in the trade negotiations already under way" (70).

There were also three new institutions created in the wake of the crisis. These included the World Food Council, which was to oversee the global food security mandate. This council was disbanded in 1993 and its functions transferred to FAO. The second was what became in 1977 the International Fund for Agricultural Development (IFAD), which still exists. The third was the Consultative Group on Food Production and Investment, which only lasted three years. So what is the way forward for the global governance of food and agriculture given what we know from recent and not so recent history of food crises? Some thirty years later, we can assess the policy outcomes from the 1974 crisis with hindsight (and I stress that these are personal observations). I focus on three aspects: productivity, trade and markets, and protecting the poor.

The Consultative Group on International Agricultural Research (CGIAR) System was built up throughout the 1970s (the first four centres combined to form the group prior to the crisis in 1971 and were joined by another nine new centres up to 1980 and four additional centres in the first years of the 1990s). Although the group was valuable in developing publicly available agricultural resources, by the beginning of the new millennium it was clear that the group was struggling (and some centres have been closed or amalgamated). The group has not been able to catalyze what is arguably the most urgent need, namely to improve technology for the rainfed agriculture that characterizes much of sub-Saharan Africa. The CGIAR is currently well along in a major organizational reform, which is much needed, since maintaining productivity growth in agriculture is essential. However, "development aid for agriculture dropped from 18 percent in 1979, to just 2.9 percent in 2006" (Båge 2008).Given the predicted adverse effects of climate change in much of sub-Saharan Africa, the urgency of the need for improved technology is ever-growing.

The world food security system reform of 1974 has not, in my opinion, been highly successful, and I am extremely skeptical that a renewed attempt with new institutions would work any better. Two of the three new institutions created after 1974 did not survive. The third one—IFAD—was not a major player in the 2008 food crisis. Although it has arguably done solid work on rural development and rural credit, this on its own was not enough to substantially improve food security.

I offer the usual economist perspective, that removing obvious market distortions is important. So, while economic theory suggests that it is reasonable to subsidize biofuel development on a temporary basis, one should review very carefully the optimal subsidies. For example, is it less harmful to subsidize ethanol from sugarcane than from maize, if sugar is grown on land that cannot readily be used for grain? Should one carefully consider the greenhouse gas emissions (which differ between crops that can generate ethanol) and take these into account when deciding what to subsidize? Not all "green" ethanol fuels are equally "green." And if, in the future, one is going to pit consumers in the Organization for Economic Co-operation and Development countries, and their demand for transportation, against poor consumers in the food-deficit countries, should there be some responsibility for ensuring better social safety nets in the developing world? These could be food-for-work programs, the currently fashionable contingent income-transfer programs, and so on. Improving safety nets is a long project, and the food crisis simply underscores its importance and urgency.

Finally, on the trade agenda, considerable progress has been made since the 1970s, despite the skepticism at that time by Hathaway. Agriculture did enter into the General Agreement on Tariffs and Trade round of the 1990s, and considerable work was done to dismantle the worst excesses of the Common Agricultural Policy and in North American policy, with their attendant disincentives to developing country agriculture. Of course, there is considerable additional work to be done. The labour-abundant poor countries in Asia have benefitted from increased market access for manufactures, and this has had considerable impact on poverty—although the benefits to Africa from increased trade are much more uncertain. I would hope that the crisis does not cause reversion to a protectionist mindset.

Given my own research agenda, I argue that there is a need to protect the nutrition of poor households in general, which is particularly important at times of crisis, and I draw two conclusions. First, interventions to improve nutrition must be cost-effective. There are over 900 million people whom the FAO classify as "hungry," who lack adequate food. Solutions that are not relatively low-cost, and cost-effective, cannot be used at scale.

Second, there is a need for continued investments in micronutrients to protect diet quality and reduce mortality and morbidity. These have been determined by the Copenhagen Consensus process to be among the top six development priorities (out of more than forty examined) on the basis of benefit-to-cost ratio, sustainability, and feasibility (Horton, Alderman, and Rivera. 2008). Investments in micronutrients, through supplements and fortification, can help to protect vulnerable populations through times when diet quality deteriorates. Although these do not solve the longer-term issues, food fortification and supplements for vulnerable groups can protect health (and in the extreme, reduce mortality rates), as longer term solutions are implemented. It is particularly unfortunate to hear of examples such as that of Senegal, which used funds for a temporary food subsidy during the crisis and then encountered difficulties in funding the Senegal Nutrition Enhancement Program (France Begin, Micronutrient Initiative, pers. comm.).

In closing I would say that the most important policy lesson of the 1974 crisis—that additional investment in agricultural development is a high priority—is even more true in responding to the 2008 crisis. Climate change is bringing additional urgency to the needs in this area, yet the rapid transition from a global food price crisis to a global financial crisis may distract our attention from the issue. I hope that we can learn from the 1974 crisis and respond better to the current crisis, so that we can prevent (or at least minimize) the next food price crisis.

Notes

1   Thinness implies that only a small proportion is traded as compared to domestic consumption.
2   The adverse weather events included drought in Australia, causing a considerably lower wheat crop, floods in Northern Europe, and a heatwave in Southern Europe, all of which disrupted production in 2007.
3   Note that the volume of food aid is slightly higher than that of cereal aid, due to the inclusion of modest amounts of non-cereal products such as skim milk powder, canned fish, etc. No source was found that had a consistent single series for either cereals only, or all food aid, from 1970 to the present.

Works Cited

Båge, Lennart (2008). "Tackling the Food Crisis: Investment, Production and Decent Work." Statement made at the International Labour Conference of the International Labour Organization, Geneva. 11 June. http://www.ifad.org/events/op/2008/ilo.htm.
Bouis, Howarth (2008). "Rising Food Prices Will Result in Severe Declines in Mineral and Vitamin Intakes of the Poor." Washington, DC: Harvest Plus.
Federal Reserve Bank (2009). Foreign Exchange Rates (monthly). Washington, DC. https://www.federalreserve.gov/econresdata/releases/statisticsdata.htm.
Federal Reserve Bank of St. Louis (2009). Historical Exchange Rates for German Mark. St. Louis. http://research.stlouisfed.org/fred2.
International Rice Research Institute (2008). "Monthly Export Price (US$/t Free on Board) of Thai Rice 5% Brokens, 1961–2008." World Rice Statistics Database. www.irri.org/science/ricestat/data/may2008/WRS2008-Table18.pdf.

Hathaway, Dale E. (1975). "The World Food Crisis—Periodic or Perpetual?" Washington, DC: International Food Policy Research Institute. http://ageconsearch.umn.edu/handle/17729.

Horton, Sue, Howard Alderman, and Juan Rivera (2008). "Copenhagen Consensus 2008 Challenge Paper: Hunger and Malnutrition." Copenhagen: Copenhagen Consensus Center.

International Feed Industry Federation/Food and Agriculture Organization (2006). "Agriculture and the Animal Feed Industry." Rome. http://www.ifif.org/files/WorldFeedOverview.ppt.

IFPRI Forum (2008). "Speculation and World Food Markets." Washington, DC: International Food Research Policy Institute. July.

Isenman, Paul (1980). "Basic Needs: The Case of Sri Lanka." *World Development* 8, no. 3: 237–58.

Kojima, Masami, and Irima Klytchnikova (2008). "Biofuels: Big Potential for Some…but Big Risks Too." Washington, DC: World Bank Institute. http://www1.worldbank.org/devoutreach/article.asp?id=506.

Lappé, Francis Moore (1971). *Diet for a Small Planet.* New York: Ballantine.

Luttrell, Clifton B. (1973). "The Russian Wheat Deal—Hindsight vs. Foresight." The Federal Reserve Bank of St. Louis Review. October. http://research.stlouisfed.org/publications/review/73/10/Russian_Oct1973.pdf.

Renewable Fuels Association (2008). Industry Statistics. Washington, DC. http://www.ethanolrfa.org/industry/statistics/.

Rogers, Paul (2008). "The World's Food Insecurity." London: OpenDemocracy. 24 March. http://www.opendemocracy.net/node/36333/pdf.

Rosegrant, Mark (2008). "Biofuels and Grain Prices: Impacts and Policy Responses." Testimony for the US Senate Committee on Homeland Security and Governmental Affairs. Washington, DC. 7 May.

Time (1974). "The World Food Crisis." *Time Magazine.* 11 November. http://www.time.com/time/magazine/article/0,9171,911503,00.html

United States Department of Agriculture (2009). "Wheat Yearbook." Washington, DC. http://www.ers.usda.gov/Data/Wheat/Yearbook?WheatYearbook.

——— (2008). "Production, Supply, and Distribution Online." Washington, DC. http://www.fas.usda.gov/psdonline.

von Braun, Joachim (2007). "The World Food Situation: New Driving Forces and Required Actions." IFPRI Food Policy Report, no. 18. Washington, DC: International Food Policy Research Institute. December.

Weinberg, Neil, and Brandon Copple (2002). "Going against the Grain." *Forbes.* 25 November. http://www.forbes.com.proxy.lib.uwaterloo.ca/forbes/2002/1125/158.html.

World Food Programme (2008). "2007: Food Aid Flows." 2008 Food Aid Monitor. Rome. http://www.wfp.org/interfais/index2.htm.

CHAPTER 4

Responding to Food Price Volatility and Vulnerability

Considering the Global Economic Context

Jennifer Clapp

In early 2008, developing countries were hard hit with rising food-import bills and widespread civil unrest sparked by sharp food-price rises. Within six months, international food prices had fallen back sharply. In some cases, such as with wheat, prices receded to levels below those of a year earlier but still higher than previous years (FAO 2008a). Understanding the range of factors contributing to price volatility is important for assessing whether it will continue, the impact on the world's poorest countries and low-income people, and how the international community should respond.

When food prices were at their height in mid-2008, a primary cause was identified in the most powerful circles: food supply was simply not keeping up with demand. This is exemplified by the comments of Jeffrey Sachs, a prominent economist and UN advisor, who explained the emergence of the crisis in May 2008 to European Union (EU) Members of Parliament in basic terms: "World demand for food has outstripped world supply" (Sachs 2008). But can the spikes in food prices be reduced to a simple equation of supply and demand? Or are more complex factors at play? As food prices climbed precipitously, and then fell quickly, a growing number of analysts pointed out that "market fundamentals" alone could not explain this price volatility. While they are important, there are broader forces at play that must be taken into consideration, particularly in shaping international governance responses to the crisis. In this chapter, I argue that broader international macroeconomic factors have played a major role not only in precipitating the recent food price volatility but also in shaping the longer-term vulnerability to price swings in the world's poorest countries. Failure to take these factors into account in the international governance response to the crisis risks the continuation of both volatility and vulnerability in the world's poorest countries.

Examining the Market Fundamentals in Recent Food Price Volatility

A survey of the main reports on rapidly rising food prices released over the past year by organizations such as the Food and Agriculture Organization of the United Nations (FAO), World Bank, Organization for Economic Co-operation and Development (OECD), the US Department of Agriculture (USDA), and the International Food Policy Research Institute (IFPRI) reveals that supply and demand fundamentals are highlighted as the most prominent causes (e.g., FAO 2008b; UN High Level Task Force 2008; World Bank 2008a; World Bank 2008b; OECD 2008; Trostle 2008; and von Braun et al. 2008). When other factors were mentioned in these reports, they were portrayed as playing a minor or supporting role. Because longer-term trends of gradually rising food prices have been forecast in recent years, it might be expected that analysis of the price spikes focus on trends already identified. But do the causes of the longer-term trends explain the sharp price rises seen in the first half of 2008?

A number of analyses pointed to rising demand for food in rapidly growing countries such as India and China (IFPRI 2008, 4; Trostle 2008, 12; OECD 2008, 2; IMF 2008, 7). At the same time, short supply was blamed on drought in Australia and bad harvests in Europe (FAO 2008b, 5; OECD 2008, 2; Trostle 2008, 2). This interpretation seemed to be confirmed by very low levels of global grain stocks. The stocks, and the "stock-to-use ratio" (the amount of stocks on hand as a percentage of overall use), were at levels not seen since the previous world food crisis in the mid 1970s (FAO 2008b, 5; OECD 2008, 2; Trostle 2008, 21).

On top of this, a rising demand for grain-based biofuels was seen to have greatly exacerbated the situation, leading to a large proportion of maize production being diverted from the food supply (OECD 2008, 2; FAO 2008b, 7–8; Rosegrant 2008). The biofuel factor affects both the demand for grain and the supply of food and, as such, has the potential to have a large influence on prices. The World Bank noted that almost all of the grain production increases experienced in the 2004–07 period went into biofuel production in the United States, thus contributing greatly to demand for grain (World Bank 2008a, 1).[1] IFPRI noted that approximately 30 percent of the food-price rises can be attributed to the diversion of maize from food markets in order to produce ethanol (Rosegrant 2008). But if biofuel production was responsible for a third of the price rises, do the other supply and demand factors make up the rest of the dramatic food price increases? This seems unlikely for several reasons.

First, in terms of demand, India and China's rising demand for food was not a sudden occurrence. Rather, it is a structural factor that has been gradually increasing over the past few decades. Moreover, these countries are self-sufficient in food and are not major buyers on global food markets (Heady and Fan 2008, 377). Second, in terms of supply, while it is true that world cereal production fell short of utilization in 2005 and 2006, there was a recovery in both 2007 and 2008 to record production levels such that stock–to-use ratios rose in 2008 (see Figure 4.1) (FAO 2008b, 6–8). This increase in global production occurred despite droughts and other bad weather that affected harvests. Moreover, the recovery in production began in 2007, yet prices continued to climb. It could be that it was not production per se but rather the stocks of grain that drove prices higher. Stocks of grain in 2007 were indeed at historically low levels. But the stock-to-use ratio in 2003 was almost as low as it was in 2007, without causing prices to jump to the same degree. In fact, the stock-to-use ratio in 2003 was lower than that experienced in 1995–96, when prices did rise sharply (see Figure 4.1). This raises questions about the extent to which grain stocks and the stock-to-use ratio automatically determine food prices.

Figure 4.1

World Cereal Stocks-to-Utilization Ratio and Cereal Price Index, 1990–2008

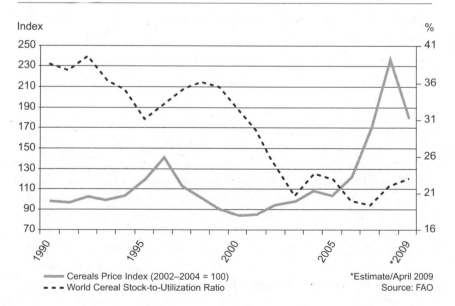

Cereals Price Index (2002–2004 = 100)
World Cereal Stock-to-Utilization Ratio

*Estimate/April 2009
Source: FAO

The FAO and others have noted that there are reasons for lower grain stock holdings that do not necessarily relate to production, such that lower stock-to-use ratios would not necessarily lead to jumps in prices. Because food prices had been low for most of the previous two decades, and storage costs high, governments moved away from storing grain. As a result, many food processors moved to a "just in time" inventory system. This move away from storing grain stocks had little to do with the amount of food produced but rather was more closely tied to the rising costs of storage (FAO 2008b, 5–6). Some have also pointed out that China, for example, had been deliberately reducing its stocks since about 2000 (Heady and Fan 2008, 381; Dawe 2009, 4–5). But because China is not a major player on world grain markets, this draw down is not considered to have had any impact on prices, and world stock-to-use ratios excluding China have been steady in recent years (ibid.).

In June 2008, the FAO noted that a significant portion of the price volatility in international food markets was beyond what could be explained by the underlying supply and demand. Futures prices for wheat, for example, were 60 percent beyond what the market fundamentals would dictate in March 2008, while prices for maize were 30 percent beyond the underlying expected value in April 2008 (FAO 2008c, 55–57). In the second half of 2008, prices fell back sharply, further contributing to the volatility. Identifying the source of this significant price volatility is of vital importance for establishing effective global governance responses to the crisis.

The International Macroeconomic Forces at Play in the Price Spikes

The fact that food prices declined quickly in the fall of 2008—just as world financial markets were collapsing—suggests that broader macroeconomic factors play a significant role in determining food prices. It may be that food prices fell in response to expected drops in demand for commodities, due to turmoil on financial markets and the ensuing global recession (FAO 2008d, 63). But this price drop signals the importance of examining the extent to which international financial factors played a role in fostering food-price volatility in the first place. The initial reports on the causes of the food crisis did mention the broader macroeconomic context, but all stressed that these factors played only a minor, supporting role. It now appears that this broader context may have had much more of a driving role in the price equation than was originally thought.

Some accounts of the food-price rises cite the weak value of the US dollar as a factor in the rise in commodity prices. The dollar's value against other currencies depreciated by 22 percent from 2002 to2007 (Abbot, Hunt, and Tyner

2008, 28; see also Timmer 2008). To avoid economic turmoil stemming from the credit crisis, the US Federal Reserve repeatedly cut interest rates in 2007–08, which kept the dollar weak against other currencies. The dollar depreciated a full 8 percent against the euro in April 2008 alone. When the US currency value falls, commodity prices in general rise. This inverse relationship, which is empirically consistent, appears to be a result of several factors, though most economists admit that the link is not fully understood (Abbott, Hunt, and Tyner 2008; Lustig 2008).

One factor that appears to be important is that as the dollar depreciates against other currencies, agricultural commodities, priced in US dollars, will see nominal price rises to account for the fact that food is priced in a currency whose value is suddenly lower. There may also be a rise in foreign demand for US grain because initially it appears to be "cheap" for those buyers whose currencies are worth more against the dollar. In addition, foreign producers, whose commodities are priced in dollars, may raise prices to compensate for the declining value of the dollar (Elliott 2008). Taking these factors into account, food-price rises in response to dollar depreciation do not necessarily signal a rise in demand for food on a global scale.

Perhaps a more important side effect of the macroeconomic context is that investment in commodities becomes particularly attractive when the value of the US dollar drops. Investors holding US dollars instead move into commodities, because they are seen to be a higher-return investment. As investors began to trade on commodity futures and exchange markets, including those markets for agricultural commodities, prices tended to rise in response to a higher demand for commodity futures contracts. Again, this does not necessarily signal a rising demand for food but rather signifies commodity speculation, or a "bubble" linked to financial investment. Peter Timmer notes this separation between real and financial factors when he states, "price formation in organized commodity markets depends on financial factors as well as 'real' supply and demand factors" (Timmer 2008, 8).

We do not know the precise impact of speculative investment in agricultural commodities on food prices. In the past it appears not to have been a concern, perhaps because most of those engaging in commodity futures markets were the direct users or producers of the commodities (known as "physical traders" or "commercial speculators"). They used the commodity futures markets as a means to hedge against their risks. In this way, commodity futures markets play an important role in mitigating risk in the agricultural sector. But in the past few years, investors with no direct interest in the commodity in its physical form (known as "non-commercial speculators") have entered futures markets in unusually large numbers. They invest with "long" positions, betting that prices will rise over time, and take their profits when only market conditions dictate (see IATP 2008).

The amounts invested in commodities generally have increased rapidly in recent years. From 2005 to March 2008, the value of commodity futures contracts doubled, to an estimated US$400 billion, climbing US$70 billion in the first three months of 2008 alone (Young 2008, 9). These large sums are mainly accounted for by large-scale investors such as sovereign wealth funds, pension funds, hedge funds, university endowments, and other institutional investors. These investors have increasingly invested in commodities via commodity index funds. These funds bundle futures contracts across a range of commodities, including agricultural commodities, into a single financial instrument based on indices such as the Dow Jones–AIG Commodity Index. Speculation in commodity index funds alone increased from US$13 billion in 2003 to US$260 billion by March 2008 (Masters 2008). Typically, agricultural commodities account for around 30 percent of the commodities in these index funds (IATP 2008).

It was not just the declining value of the dollar that attracted these noncommercial speculators. Loopholes in the regulatory framework in the United States have also encouraged this activity. These large-scale investors were effectively exempted from "position limits" when they "swap" futures contracts through financial institutions that hedge what are known as "over-the-counter" swap transactions. In effect, large-scale investors can invest huge sums in commodities futures, but they do so via large Wall Street banks that act as middlemen. The problem is that financial investors tend to speculate on the markets in order to make money, and they are not interested in the physical commodity they purchase. They go into and out of commodity markets largely in reaction to market algorithms based on broader macroeconomic conditions (IATP 2008). It is difficult to know which comes first, the speculation or the higher prices, as they are tightly linked. The result, however, is the same: large swings in the price of agricultural commodities, including basic food staples.

The general rise in commodity prices in this period was also linked to rapidly rising prices for oil, which in turn had an influence on food prices. The price of a barrel of oil reached nearly US$150 in mid-2008 before it began to drop significantly when the financial markets collapsed (the price of oil as of March 2009 hovered around US$50 per barrel). Farm inputs such as pesticides and fertilizers are petroleum-based products, and as oil prices rose in the first half of 2008, farm costs also rose. Perhaps more importantly, the rise in oil prices also fuelled investment in biofuels (which had suddenly became much more economically viable), also driving up grain prices.

As food prices rose precipitously in the early months of 2008, a number of developing countries, including Vietnam, India, China, Argentina, and Egypt, began to impose trade restrictions on agricultural exports (World Bank 2008b, 2; von Braun et al. 2008, 5). The aim was not only to keep foodstuffs at home but also to insulate their economies from high and rising prices of food on international markets. These actions were largely a response to external conditions, including speculation-induced price spikes. Although this strategy can help with food availability and prices at home, it can seriously exacerbate the price situation on international markets. The largest food price spikes for wheat and rice, for example, occurred on days when export restrictions were announced in major food-exporting developing countries.

Most reports outlining key causes of food-price spikes did attribute a supporting role to these broader international economic forces but put much more blame on basic supply and demand factors. Of the international forces mentioned, the export restrictions put in place by developing countries were seen to be much more problematic than the effects of a declining dollar and commodity speculation. The World Bank, for example, stated that "although the empirical evidence is scarce, the prevailing consensus among market analysts is that fundamentals and policy decisions are the key drivers of food-price rises, rather than speculative activity" (World Bank 2008b, 2). IFPRI's policy brief similarly noted that while speculation played a supporting role, it was more of a symptom of food-price rises, rather than a cause (von Braun et al. 2008).

Since the air was let out of the international food-price bubble when the financial markets collapsed in the fall of 2008, there has been a growing acknowledgement of the importance of commodity market speculation even among skeptics (Timmer 2008, 7). The worry now, however, is that the financial collapse is drying up sources of credit, a great risk for developing-world farmers (FAO 2008d, 63-64),who pull back production when unable to finance inputs and hedge their risks on futures markets. Rapidly falling agricultural commodity prices are particularly harmful for indebted farmers, a category that includes most of the world's farmers. The effects of this situation could make the food price bubble of 2007–08 seem mild in comparison.

Crisis on Top of a Crisis: Global Economic Contributions to Vulnerability in Developing Countries

Agricultural price volatility is particularly problematic for the world's poorest countries, which are typically agricultural-based and dependent on food imports. The FAO lists eighty-two countries as "Low-Income Food-Deficit Countries" (LIFDCs), which are especially vulnerable to sharp movements in international food prices.[2] According to the FAO, the least developed countries—most of which are in the category of LIFDC—were net agricultural exporters in the 1960s. Today, as a group, they are net agricultural importers (see Figure 4.2). How did this situation arise? This is a debated question, and to focus only on domestic factors within these countries would provide an incomplete picture. There are a number of broader international factors that have contributed to the import dependence. These various forces are complex and wide-ranging, and only a brief overview is provided here.

Industrialized country agricultural trade policy is widely seen to have had a negative impact on developing country agriculture. Developing countries have complained in international trade negotiations that their incentives for increasing domestic production are harmed by agricultural subsidies of over US$300 billion per year in the industrialized countries. These subsidies have been blamed for encouraging the dumping of cheap agricultural products on world markets, depressing world agricultural commodity prices for most of the past thirty years (Oxfam 2005; Murphy, Lilliston, and Lake 2005). In addition, industrialized countries have restricted access to their markets for products from developing countries through tariff peaks and tariff escalation practices. Most developing countries liberalized their trade policies under programs of structural adjustment in the 1980s and 1990s and cannot afford to subsidize their own farmers to counteract the trade practices of the industrialized countries. This highly uneven playing field has been identified by many analysts as a key cause of reduced incentives for agricultural production in developing countries in the past two decades (see Khor 2005; Weis 2007).

Figure 4.2

Agricultural Trade Balance of Least Developed Countries, 1961–2006

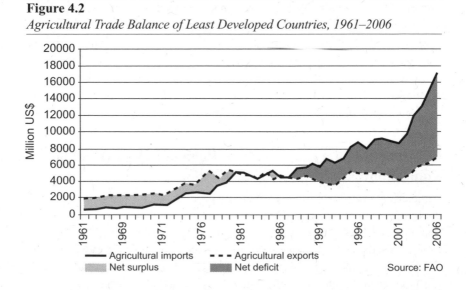

The challenges that industrialized country trade policies pose for developing country agricultural sectors were recognized at the launch of the Doha Round of World Trade Organization (WTO) trade talks in 2001. The WTO Agreement on Agriculture, which brought agriculture under international trade rules with the completion of the Uruguay Round in 1994, was to be renegotiated under the Doha Round to rectify these imbalances. After over eight years of contentious negotiations over agriculture, both among rich countries and between rich and poor countries, however, a deal has not yet been reached. The industrialized countries—mainly the United States and the EU—have been reluctant to reduce their domestic support subsidies to their own farmers, while pushing hard for greater market access in developing countries (Clapp 2006; Rosset 2006). The most recent collapse of the talks was over the details of a safeguard mechanism that developing countries insist on putting in place to protect themselves from import surges of low-priced agricultural imports that threaten their own farmers' livelihoods.

As international trade policies dampened incentives for agricultural improvements in the world's poorest countries, investment in agriculture in those same countries also declined over the past thirty years. World Bank lending for agriculture, for example, has declined from 30 percent of its overall lending in 1980 to just 12 percent in 2008, while the percentage of official development assistance earmarked for agriculture dropped from 18 percent to 3 percent in the past two decades (World Bank 2008a, 8; UN High Level Task Force 2008, 8). Developing country governments also failed to invest in this sector, as high levels of international debt strained their budgets. This drop in agricultural investment coincided with historically low world food prices. With cheap food imports available over a period of some thirty years, incentives to improve their own agricultural systems were weak at best.

For the last thirty years, food crises have been addressed largely via food aid, which has provided a stopgap rather than a viable long-term investment for the food sector. In-kind food aid has been problematic, however, as it can create additional disincentives for local production and can also lead to market distortions (Barrett and Maxwell 2005). Most donors have moved to cash-based food aid provision, including the EU since the mid-1990s and Canada and Australia more recently. But the United States, by far the world's largest donor of food aid—accounting for nearly 50 percent of all food aid—maintains a nearly 100 percent tied food-aid policy, meaning that the aid must be sourced in the donor country (OECD 2005).

Increasing corporate concentration and control in the global food and agriculture system has also brought difficulties for many developing countries

(Clapp and Fuchs 2009; Heffernan 2000). A small handful of corporations dominate the markets for nearly every aspect of the global food system, from seeds to commodity trade to processing and retailing. At the retail end of the spectrum, privately set quality standards established by market-dominating retailers have created a situation in which it is difficult for small-scale farmers in developing countries to sell their products to international retailers, further reducing incentives for increased production (Hatanaka, Bain, and Busch 2005; McMichael and Friedmann 2007).

Weak agricultural performance and growing dependence on food aid and other imported food in the world's poorest developing countries have resulted in rural poverty and heightened vulnerability to price shocks. In such a context, suddenly higher food prices were not a "boon" to rural economies in the developing world.

Solutions to the Crisis Must Respond to Root Causes of Volatility and Vulnerability

Given the vulnerability of developing countries in the current global food economy, effective global governance responses are imperative. The price rises in the first half of 2008 led a number of key international organizations to put forward policy proposals to address both the short- and longer-term challenges. The UN Secretary-General, Ban Ki Moon, established the High Level Task Force on the Global Food Crisis in April 2008, and its initial report was circulated at the High Level Conference on Food Security held in early June 2008 (UN High Level Task Force 2008). The G8 met in July 2008 and released a Statement on Global Food Security (G8 2008). Other organizations provided input into these discussions, resulting in policy recommendations from the World Bank (2008b), OECD (2008), the USDA (Trostle 2008), and IFPRI (von Braun et al. 2008).

These policy documents are consistent on four key proposals for an international governance response to the food crisis: provide emergency aid and loans to meet short-term food needs, increase supply via agricultural investment in developing countries, temper demand by rethinking biofuel policies, and reduce supply bottlenecks by improving the functioning of international food trade. These policy responses follow directly from the interpretation of the crisis by these same institutions: that the problem is largely one of supply and demand fundamentals and exacerbated mainly by export restrictions imposed by developing countries. Without taking away from the importance of these recommendations for improving agriculture in developing countries, it is clear that they do not fully take into account the international economic forces that have contributed to both the volatility of food prices and the vulnerability of these countries.

Provide Emergency Aid and Loans

All of the policy documents noted above recommend policy measures to provide emergency assistance to meet short-term food needs. This includes increased funding for food aid programs as well as emergency balance of payments loans to enable food-deficit countries to pay for food imports. These measures are certainly welcome, because in early 2008 the World Food Programme found itself short by over US$700 million. This money was largely forthcoming from donor governments in the height of the price rises, but the aid from the United States was still largely tied to in-kind aid in the form of commodities, rather than cash. The latest Farm Bill in the United States—passed in the midst of the food price crisis—includes a modest pilot project for cash food aid for local and regional purchase. However, given the extreme conditions, the international community could have put much more pressure on the United States to reform its food aid policies more thoroughly in ways that reduce market distortions and hamper domestic production incentives in developing countries.

Increase Supply through Agricultural Investment

The promotion of agricultural production in developing countries via increased aid and loans for agricultural inputs, infrastructure, and research was a strong common thread through the policy documents. All of the proposals included specific mention of the role of science and technology in agricultural research. The OECD and the G8 Leaders Statement on Global Food Security both explicitly mention the need to promote agricultural biotechnology as a way to increase food production in developing countries. IFPRI and the High Level Statement were the only two reports to explicitly call for agricultural investment to be sensitive to the needs of promoting ecologically sound agriculture. None of the policies recommended in these various documents, however, mentions the International Assessment of Agricultural Knowledge, Science and Technology for Development (IAASTD) report, the summary of which was released in April 2008, when food prices were experiencing serious spikes (IAASTD 2008; IAASTD 2009). This report, the result of a multi-year and multi-institution- and government-supported effort, was highly skeptical of the benefits of GMOs in agriculture and promoted small-scale sustainable agriculture as opposed to large scale commercial agriculture. The fact that this report was completely ignored, even by the very institutions that initially sponsored the process, speaks volumes.[3]

Temper Demand by Rethinking Biofuel Policies

Nearly all of the policy documents call for tempering demand for grain by rethinking biofuel policies. The USDA was an exception, downplaying the significance of this factor. But all stop short of calling for mandatory measures, and instead call for further research on the issue. IFPRI's policy brief gives suggestions about the types of policies that might be put into place, such as imposing a moratorium on biofuels based on grains and oilseeds until prices drop and increasing support to non-food-based biofuels (von Braun et al. 2008). The report of the High Level Task Force suggested a possible international biofuel consensus as a way to ensure broad-based support for less-damaging biofuel policies (UN High Level Task Force 2008, 24–25).

Improve International Agricultural Markets

All the policy documents call for an end to export bans in developing countries and a swift completion of the Doha Round. There is no doubt that export restrictions exacerbated the price-spike situation, but it is not clear whether they were the cause of initial sharp price rises. Developing countries were probably responding to the international context, in which global prices for foodstuffs were rising quickly, and not to their own domestic production situations. Completion of the Doha Round will only be beneficial if it results in drastically reduced subsidies in the industrialized countries and safeguards for developing countries from surges of cheap imports, enabling them to protect farmer livelihoods. It is not worth the risk of agreeing to another unbalanced deal just for the sake of concluding the Doha Round. It is important that the role of the international context, in which industrialized countries have long affected developing country agriculture through their own protectionist agricultural policies, be recognized and redressed in international agricultural trade rules.

What Was Missing

Conspicuously missing from the various documents is a set of policy proposals directed at putting strict regulations on commodity markets to limit speculation by non-commercial market participants. Only IFPRI and the UN High Level Task Force report note that the issue could use further study and that regulatory measures might be considered. But IFPRI warned of the risk of over-regulation and called for measures to be "market-oriented" (von Braun et al. 2008, 9). This contrasts sharply with recommendations from more critical organizations. The Institute for Agriculture and Trade Policy, for example, has called for multilateral efforts to reduce agricultural commodity speculation in order to dissuade investors from "exchange shopping" (IATP 2008, 10).

Also missing from documents reviewed here is much mention of the role of corporate concentration in the global food system.[4] The World Bank does call for increased private-sector investment in agribusiness as a way to encourage more production in developing countries. But little has been said about the ways in which corporate concentration might be hindering developing country production incentives and retail market access for their agricultural goods. As global corporations come to dominate most aspects of the global agricultural supply chain, it is important to carefully consider the impact of these developments on developing country agricultural systems.

Conclusion

Unexpected food-price spikes in the international arena have the capacity to wreak havoc by deepening poverty, hunger, and political unrest in developing countries. In this context, global governance responses to the situation should target not just the "market fundamentals" of supply and demand for food, but also the broader volatility and vulnerability aspects of the crisis. Because they have prioritized basic supply and demand issues as causes of the crisis over both the short- and longer-term international macroeconomic forces affecting developing country agriculture, the policy proposals put forward by the key international institutional players are incomplete.

The international economic context must be taken into account if the proposed policy measures are to be successful in avoiding future crises. What is missing from present policy proposals is substantial movement toward a *global* economic framework that facilitates agricultural development in the South. There is a need to set rules that impact the North as much as the South—such as trade rules that actually create policy space for developing countries and rectify longstanding imbalances caused by agricultural subsidies in the industrialized countries. Also missing is serious movement toward regulation on agricultural commodity speculation—so that international prices are not subject to sharp swings to which developing countries must react and from which they must protect themselves.

Notes

This chapter is a revised version of Jennifer Clapp (2009). "Food Price Volatility and Vulnerability in the Global South: Considering the Global Economic Context." *Third World Quarterly* 30, no. 6. Portions of that article are reprinted here with permission.

1   The USDA analysis, it should be noted, downplays the significance of the biofuel factor (Trostle 2008, 15–18).

2   This group includes most countries in sub-Saharan Africa, much of Asia including India, China, the Philippines, Bangladesh, Pakistan, and Indonesia, and several Central American and Caribbean countries (FAO 2008e).

3   Please see Marcia Ishii-Eiteman's chapter in this volume for further analysis of the conclusions drawn by the IAASTD report.

4   An earlier IFPRI document, however, does mention this factor (von Braun 2007).

Works Cited

Abbot, Philip, Christopher Hunt, and Wallace Tyner (2008). "What's Driving Food Prices?" Oak Brook: Farm Foundation.

Barrett, Christopher, and Daniel Maxwell (2005). *Food Aid after Fifty Years: Recasting Its Role*. London: Routledge.

Clapp, Jennifer (2006). "WTO Agriculture Negotiations: Implications for the Global South." *Third World Quarterly* 27, no. 4: 563–77.

Clapp, Jennifer, and Doris Fuchs, eds. (2009). *Corporate Power in Global Agrifood Governance*. Cambridge: MIT Press.

Dawe, David (2009). "The Unimportance of 'Low' World Grain Stocks for Recent World Price Increases," ESA Working Paper no. 09-01 (February). Rome: FAO.

Elliott, Larry (2008). "Against the Grain: Weak Dollar Hits the Poor." *The Guardian*. 21 April.

Food and Agriculture Organization of the United Nations (2008a) "Crop Prospects and Food Situation," no. 4 (October). Rome. http://www.fao.org/docrep/011/ai473e/ai473e00.htm.

———— (2008b). "Soaring Food Prices: Facts, Perspectives, Impacts and Actions Required." HLC/08/INF/1. Rome. ftp://ftp.fao.org/docrep/fao/meeting/013/k2414e.pdf.

———— (2008c). "Food Outlook." (June). Rome. ftp://ftp.fao.org/docrep/fao/010/ai466e/ai466e00.pdf.

———— (2008d). "Food Outlook." (November). Rome. ftp://ftp.fao.org/docrep/fao/011/ai474e/ai474e00.pdf.

———— (2008e). "Low-Income Food-Deficit Countries (LIFDC)." Rome. http://www.fao.org/countryprofiles/lifdc.asp?lang=en.

Hatanaka, Maki, Carmen Bain, and Lawrence Busch (2005). "Third Party Certification in the Global Agrifood System." *Food Policy* 30, no. 3: 354–69.

Heady, Derek, and Shenggen Fan (2008). "Anatomy of a Crisis: The Causes and Consequences of Surging Food Prices." *Agricultural Economics* 30: 375–91.

Heffernan, William (2000). "Concentration of Ownership and Control in Agriculture." In *Hungry for Profit. The Agribusiness Threat to Farmers, Food and the Environment*, ed. Fred Magdoff, John Bellamy Foster, and Frederick H. Buttel, 61–75. New York: Monthly Review Press.

Institute for Agriculture and Trade Policy (2008). "Commodities Market Speculation: The Risk to Food Security and Agriculture." Minneapolis. http://www.iatp.org/tradeobservatory/library.cfm?refID=104414.

International Assessment of Agricultural Knowledge, Science and Technology for Development (2008). "Executive Summary of the Synthesis Report." Washington, DC: Island Press. http://www.agassessment.org/docs/IAASTD_EXEC_SUMMARY_JAN_2008.pdf.

International Monetary Fund (2008). "Food and Fuel Prices—Recent Development, Macroeconomic Impact and Policy Responses." Washington, DC. 30 June. http://www.imf.org/external/np/pp/eng/2008/063008.pdf.

Khor, Martin (2005). *The Commodities Crisis and the Global Trade in Agriculture: Problems and Proposals*. Malaysia: Third World Network.

Lustig, Nora (2008). "Thought for Food: The Challenges of Coping with Soaring Food Prices." Working Paper no. 155. Washington, DC: Center for Global Development.

Masters, Michael (2008). Testimony before US Senate Committee on Homeland Security and Governmental Affairs. Washington, DC. 20 May.

McMichael, Philip, and Harriet Friedmann (2007). "Situating the Retail Revolution." In *Supermakets and Agri-food Supply Chains: Transformations in the Production and Consumption of Foods*, ed. David Burch and Geoffrey Lawrence, 154–72. Cheltenham: Edward Elgar.

Murphy, Sophia, Ben Lilliston, and Mary Beth Lake (2005). "WTO Agreement on Agriculture: A Decade of Dumping." Minneapolis: Institute for Agriculture and Trade Policy.

Organization for Economic Co-operation and Development (2008). "Rising Agricultural Prices: Causes, Consequences and Responses." Policy Brief. *OECD Observer.* August.

——— (2005). "The Development Effectiveness of Food Aid: Does Tying Matter?" Paris: OECD.

Oxfam (2005). "A Round for Free: How Rich Countries are Getting a Free Ride on Agricultural Subsidies at the WTO." Oxfam Briefing Paper no. 76. Oxford: Oxfam. June. http://www.maketradefair.com/en/assets/english/aroundforfree.pdf.

Rosegrant, Mark (2008). "Biofuels and Grain Prices: Impacts and Policy Responses." Testimony to the US Senate Committee on Homeland Security and Governmental Affairs. Washington, DC: International Food Policy Research Institute. 7 May.

Rosset, Peter (2006). *Food is Different: Why the WTO Should Get Out of Agriculture.* London: Zed.

Sachs, Jeffrey (2008). Speech to the European Parliament Committee on Development. Brussels. 5 May.

Timmer, C. Peter (2008). "The Causes of High Food Prices." Asian Development Bank Working Paper no. 128. Manila: Asian Development Bank. http://www.adb.org/Documents/Working-Papers/2008/Economics-WP128.pdf.

Trostle, Ronald (2008). "Global Agricultural Supply and Demand: Factors Contributing to the Recent Increase in Food Commodity Prices." Washington, DC: US Department of Agriculture.

United Nations High Level Task Force on the Global Food Crisis (2008). "Elements of a Comprehensive Framework for Action." New York: United Nations. June.

von Braun, Joachim (2007). "The World Food Situation: New Driving Forces and Actions Required." IFPRI Food Policy Report. Washington, DC: International Food Policy Research Institute. http://www.ifpri.org/pubs/fpr/pr18.pdf.

von Braun, Joachim, et al. (2008). "High Food Prices: The What, Who and How of Proposed Policy Actions." IFPRI Policy Brief. Washington, DC: International Food Policy Research Institute. http://www.ifpri.org/PRESSREL/2008/pressrel 20080516.pdf.

Weis, Tony (2007). *The Global Food Economy: The Battle for the Future of Farming.* London: Zed.

World Bank (2008a). "Rising Food Prices: Policy Options and World Bank Response." Washington, DC: World Bank. http://siteresources.worldbank.org/NEWS/Resources/risingfoodprices_backgroundnote_apr08.pdf.

——— (2008b). "Double Jeopardy: Responding to High Food and Fuel Prices." Washington, DC. 2 July. http://www-wds.worldbank.org/external/default/WDSContentServer/WDSP/IB/2008/08/08/000333038_20080808113520/Rendered/PDF/449510WP0Box321ummit1paper01PUBLIC1.pdf.

Young, John (2008). "Speculation and World Food Markets." *IFPRI Forum.* July.

CHAPTER 5

US Biofuels Policy and the Global Food Price Crisis

A Survey of the Issues

Kimberly Ann Elliott

"The RFS [renewable fuel standard] remains an important tool in our ongoing efforts to reduce America's greenhouse gas emissions and lessen our dependence on foreign oil, in aggressive yet practical ways." – Environmental Protection Agency administrator Stephen L. Johnson announcing his decision to deny a waiver of the mandated minimum level of ethanol in gasoline, August 7, 2008 (US EPA 2008).

Just months before riots spurred by high food prices broke out in developing countries around the world, the US Congress passed, and former president George W. Bush signed, legislation aimed at promoting energy independence, including a sharply increased minimum level of "renewable fuels" in gasoline. With current technologies, "renewable" means mainly corn-based ethanol in the United States, sugar-based ethanol in Brazil, and oilseed- or palm-oil-based biodiesel in the European Union (EU). Production of these commodities for fuel competes with food production, either directly or by diverting acreage from food crops. As the energy bill mandate was being finalized, Congress was also debating farm legislation that included an extension of the US$0.54 per gallon tariff on imported ethanol and modestly reduced the tax credit for refiners using ethanol, from US$0.51 to US$0.46 per gallon.

Eight months later, with season-average corn prices projected to be more than 50 percent higher than a year earlier—and with the World Bank estimating that 100 million people in developing countries would be pushed back into poverty—US Environmental Protection Agency (EPA) administrator Steve Johnson made the announcement affirming the Bush administration's support for biofuel subsidies. The European Union has similar tax and regulatory policies promoting the use of biofuels. Biofuel advocates usually cite one or more of the same rationales as Johnson—improving energy security by reducing dependence on foreign sources of oil, reducing greenhouse gas emissions, or boosting rural livelihoods. But with food prices surging, these policies are attracting greater scrutiny.

Skepticism regarding the security and environmental benefits of the current generation of food-based biofuels is not new. But the critiques became sharper and louder with the acceleration of food-price increases in the first half of 2008. Moreover, while past research raised serious questions about the overall climate change benefits from corn-based ethanol, new research that takes into account deforestation and other land-use changes concludes that the current generation of food-based biofuels is more likely to contribute to global warming than mitigate it. Climate change, in turn, is expected to threaten agricultural sustainability in tropical areas, especially sub-Saharan Africa, making food insecurity an even more serious problem in the future (Cline 2007).

This chapter examines the role that biofuels, especially corn-based ethanol, and policies promoting them, might be playing in influencing food prices. The chapter puts forward three main arguments. First, demand for ethanol is the most significant factor in the rise in corn prices. It is also important for soybean prices because many farmers have interrupted their normal practice of rotating acres between corn and soybeans. In addition, EU policies promoting biodiesel raise demand and prices for oilseeds and palm oil. Second, although the magnitude of any spillover to other grains and food products is harder to pin down, biofuels have played a role by diverting production from and consumption to alternative crops. Finally, corn ethanol is not making a significant contribution to the energy security and environmental goals set for it and the policies promoting it are costly to taxpayers and the environment.

Even though some proponents concede the limitations, they argue that government support for corn ethanol is necessary as a "bridge" to the next generation of potentially more efficient and environmentally effective biofuels, including those made from algae, agricultural waste, or from grasses or jatropha (a tropical shrub) grown on marginal land not suitable for food crops. But development of viable alternatives is slowed rather than accelerated by diverting resources to corn ethanol and creating a production and distribution infrastructure that may not be transferable or in the right place when the next generation of biofuels can be commercialized. Moreover, sugar ethanol from Brazil, which has greater net energy and environmental benefits (as long as it does not contribute to further deforestation in the Amazon region) is available now but is discouraged by an import tax.

Critics should not overestimate the degree to which eliminating biofuel subsidies would stem rising food prices when oil prices rise again, however. High gasoline prices boost the demand for alternatives and make ethanol an economically competitive alternative. With the infrastructure in place thanks to past subsidies, the link between energy and food prices will kick in again when economic recovery boosts the demand for gasoline. Reducing subsidies last year when prices were high would have been helpful this year for removing incentives to maintain production capacity in the face of falling prices.

Unfortunately, the tax credit and tariff were extended in the 2008 US Farm Bill, and changes to those policies require additional action by the US Congress. The EPA has the authority to waive all or part of the mandate for up to one year at any time but declined to do so in summer 2008 despite a request from Texas Governor Rick Perry, who was concerned about the health of his state's livestock industry. A waiver then would likely have provided little more than symbolic relief. Nonetheless, because the mandate was not binding given high gasoline prices, it is propping up production in the midst of recession and plummeting oil prices.

Factors behind Food-price Increases

Longer-run trends in both food supply and demand contributed in recent years to global grain stocks reaching historic lows. Much of the global adjustment was due to China drawing down stocks that had reached unusually high levels in the late 1990s, but US stocks have also recently fallen to low levels (Schneph 2008). Low stocks were not the trigger for the recent price spikes, but they set the stage for them. Tight supplies meant there was very little cushion to absorb sudden changes in demand, such as for biofuels, and cyclical supply shocks, such as the prolonged drought in Australia. These factors amplified the price effects, which were further exacerbated by macroeconomic trends and shocks outside agriculture, including the declining dollar, which boosted demand for US exports, the popping of the real estate bubble, and inflationary expectations, which drove investors and speculators into commodity futures markets as a hedge (Trostle 2008).[1]

The longer-run trends affecting prices include rising demand in large developing countries, especially for meat and dairy products (several pounds of grain are required for each pound of meat produced).[2] A result of growth and rising incomes, the rise in demand is both welcome and here to stay. On the supply side, investments in agriculture have been declining for more than two decades, especially in developing countries. This trend is reversible, albeit limited by available land and water. In addition, subsidies and trade protection provided by rich countries to their farmers, which averaged one-third of gross farm receipts from the mid-1980s to the early 2000s, pushed down world prices and discouraged increased investments in agriculture in developing countries (Elliott 2006). These factors contributed to stocks of corn, rice, and wheat that peaked in the late 1990s at over 500 million metric tons and then fell to 300 million metric tons, just over 10 percent of needs (Figure 5.1).

In addition to these long-run trends, sharply rising energy prices both increased the demand for alternative fuels, such as ethanol, and raised production costs. Demand and prices for corn and vegetable oils, for example, rose sharply as fuel uses competed for limited supplies. The price of fertilizer, which is energy-intensive in production, ranged two to four times higher in May 2008 than the average for 2006, while the price of phosphate rock climbed sevenfold (World Bank 2009). The declining dollar also contributed by dampening price increases in foreign currency terms, which increased demand for US exports and further boosted the US dollar price.[3] Adverse weather, especially in key wheat-producing areas, also contributed to unusually tight supplies. Over the longer run, climate change is projected to exacerbate drought in some areas, especially sub-Saharan Africa, and floods in others (Cline 2007). Table 5.1 summarizes the factors contributing to the food price crisis.

Figure 5.1
World Grain Stocks to Use

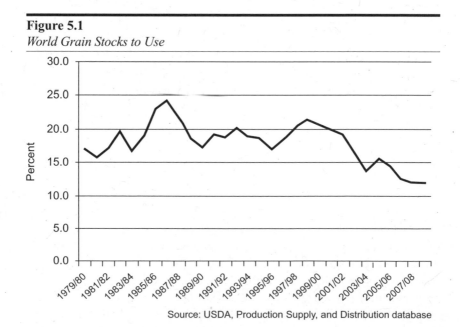

Source: USDA, Production Supply, and Distribution database

Table 5.1

An Illustration of Factors in the Food Price Crisis

	Demand Side	**Supply Side**
Long run	Growth, rising incomes in developing countries leading to increased demand for meat, dairy products and indirect demand for grains	Inadequate investments in research and development, infrastructure, and extension services to increase productivity

Effect of long-run trends: Demand growth > Supply growth = Declining stocks

Recent, emerging	Biofuels demand	Rising energy, other costs
Short run, cyclical	Financial speculation?	Adverse weather
		Bad policies, including export restrictions, hoarding and pre-emptive buying, price controls, untargeted subsidies

Biofuels and Food Prices

Figure 5.2 illustrates the links between feed corn prices and the share of corn production in the United States going into ethanol, and the production of ethanol. There is not much correlation between the corn price and ethanol production until 2005–06, when ethanol production, the share of US corn production going to ethanol, and corn prices all surged upward.

Figure 5.2

Corn Markets and Ethanol Production

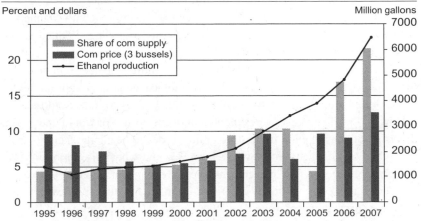

Source: Renewable Fuels Association, Industry Statistics, online, USDA, Economic Research Servcie, Feedgrains Database, online

Most analyses conclude that increased demand for ethanol has been the major factor in rising corn prices (see, for example, Yacobucci and Schnepf 2007; Collins 2008). But ethanol supporters argue that ethanol demand has little or nothing to do with the recent food-price increases because people do not eat feed (yellow) corn. Although this is true, people do indirectly eat feed corn when they eat meat, especially poultry, dairy products, and eggs. Prices of the latter two items are projected to rise roughly 50 percent in the United States this year (Yacobucci and Schnepf 2007). The explanation of price increases in staple grains that people eat, including white corn, wheat, and rice, is more complicated.

There are two main channels through which increased demand for corn-based ethanol might affect other grain prices:

- shifting acreage from production of other crops to corn, thereby reducing supplies and raising prices for the competing crops; and
- shifting consumption, both by people and animals, from corn to other staple grains.

In the United States, many farmers rotate land between corn and soybeans to maintain soil quality and yields. Wheat- and corn-growing areas overlap only along the fringes and relatively little substitution of corn for wheat in production is expected. While the number of acres planted in wheat did decline slightly from 2003 to 2005 as corn acreage increased, the big surge in acres planted in corn (when ethanol demand surged in 2006–07) came out of soybean acres (Figure 5.3). But the sharp rise in soybean prices that occurred induced farmers to reverse some of the production shift in the United States.

Biofuels thus contributed to rising oilseed prices through both channels— reducing supply by diverting acreage from soybeans to corn in the United States, and by increasing demand for oilseeds for biodiesel in Europe.

It is more difficult to identify a link between biofuel demand and the surge in rice prices early in 2008. Land and climatological conditions appropriate for growing rice are generally not suitable for other crops, so one would expect relatively little diversion of acres planted in rice to corn. Nor would one expect diversion of consumption from other grains to rice to be a large factor—aside from the Indian subcontinent where both rice and wheat are important (see below). Only about a third of rice consumption occurs outside Asia, where rice accounts for 50 percent of daily calories consumed. In Latin America and Sub-Saharan Africa, however, roughly 15 percent of calories are provided by corn versus 8 to 9 percent from rice, so some switching in reaction to high corn prices is possible (Table 5.2). Wheat makes up another 7 to 13 percent of daily calories consumed in Sub-Saharan Africa and Latin America, respectively, and

Figure 5.3

Acreage Planted by Commodity

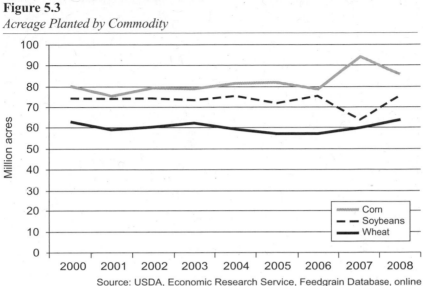

Source: USDA, Economic Research Service, Feedgrain Database, online

although a switch to more wheat consumption could be factor, it does not seem it would be a large one.

Plantings of wheat were up sharply in 2008 in response to high prices, and production was expected to be nearly 10 percent above the average for the past three years (USDA 2008). This in turn contributed to a significant softening in wheat prices. Because the wheat price rise was somewhat ahead of that for the other grains, it is not clear that the wheat price story can be explained by consumption switching from corn. Moreover, as stated above, there was relatively little production diversion, at least in the United States, which accounts for roughly 10 percent of global production and a quarter of exports. There were, however, a number of adverse weather events in key wheat-producing areas, including Australia and Ukraine.

In addition, as Slayton and Timmer explain, weather-related damage to the local wheat crop in 2006 and the desire to avoid expensive wheat imports were behind the Indian Government's decision in late 2007 to ban exports of non-basmati rice in order to ensure adequate domestic food supplies (Slayton and Timmer 2008). That, in turn, triggered export bans and hoarding elsewhere in Asia, which was a major factor in the rice price spike in early 2008. Mitchell attributes the Indian decision to avoid expensive imports to tight supplies and high prices that, he argues, were caused to a significant degree by increased demand for biofuels (Mitchell 2008). Thus, the rice panic story is related, in his view, to biofuels production, and the sharp price rise (up to February 2008

Table 5.2

Sources of Daily Calories in Developing Countries, 2002

	Sub-Saharan Africa		Latin America, Caribbean		South Asia		East, SE Asia	
	Percent of daily calories	Import share	Percent of daily calories	Import share	Percent of daily calories	Import share	Percent of daily calories	Import share
Cereals, starchy roots	66	21	40	31	63	2	64	25
Wheat	7	77	13	62	21	3	6	105
Rice	8	42	9	16	35	negl.	49	5
Maize	15	9	14	18	2	2	5	38
Sorghum, millet	14	1	negl.	33	3	negl.	negl.	negl.
Starchy roots	20	negl.	4	2	2	negl.	4	6
Vegetable oils	8	32	10	33	9	52	7	32
Animal products	6	16	20	11	8	1	9	26

when his analysis stops) is included in his assessment that "three-quarters of the 140 percent actual [food price] increase was due to biofuels and the related consequences of low grain stocks, large land use shifts, speculative activity, and export bans" (ibid., 1).

In sum, most analysts have concluded that the increase in ethanol production is the major cause of rising corn prices since 2005 (Yacobucci and Schnepf 2007; Collins 2008). Along with some shifts in plantings, weather seems to be an important factor in the case of wheat; rising prices for rice were caused mainly by speculative hoarding and panic in Asia. Mark Rosegrant of the International Food Policy Research Institute (IFPRI) estimates that from 2000 to 2007, biofuels caused 39 percent of the rise in corn prices, 21 percent for rice (keeping in mind that much of the rice price surge occurred in 2008), and 22 percent for wheat. He estimates that biofuels account for 30 percent of the overall weighted average increase in grain prices over that period (Rosegrant 2008). The Organization for Economic Co-operation and Development (OECD) similarly blames biofuels for a third of the projected increase in cereal and oilseed prices over the next decade, relative to the average level over the past decade (Boonekamp 2008). Mitchell's estimate that 75 percent of the food-price increase is due to biofuels and related supply effects is very much on the high side (Mitchell 2008).

Energy, Food Prices, and US Biofuel Policy

Rising energy prices affect food prices on both the supply and demand sides. Rising oil and natural gas prices raise the costs of producing food and transporting it to markets. Agriculture in rich countries, where commodities are produced using diesel-powered machines and large amounts of fertilizer and pesticides, which in turn are energy-intensive in their production, is particularly affected by rising energy costs. And, on the demand side, rising gasoline prices make ethanol and other biofuels economically attractive. Figure 5.4 illustrates the correlation between gasoline prices and ethanol production. Oil and other commodity prices dropped sharply in the second half of 2008, but gasoline prices started rising again in Spring 2009 despite the sharp contraction in economic growth. Most projections are for commodity prices to resume rising when the economic recovery begins.

The chart also indicates some of the effects of government policies and their impact on ethanol production. The US Congress approved subsidies for adding ethanol to gasoline following the first oil-price shock (and commodity boom) in the 1970s. Today, there are a numerous federal and state subsidies for biofuels, but the most important are a credit against the excise tax on gasoline, an import

Figure 5.4
Gasoline Prices and Ethanol Production

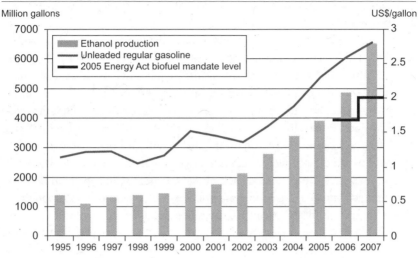

Source: Renewable Fuels Association, Industry Statistics, online

duty designed to offset the benefits of the tax credit for foreign-produced ethanol, and a mandate setting minimum levels of biofuels for use in transportation fuels and home heating oil. The credit against the federal excise tax on gasoline has changed in detail, but it has fluctuated around US$0.50 per gallon for many years. It was reduced to US$0.46 per gallon in the 2008 farm bill in years in which ethanol production is above 7.5 billion gallons, which is well below the 2009 mandate level of 10.5 billion gallons. In addition, there is a US$0.54 per gallon tariff, primarily to discourage imports of sugar-based ethanol from Brazil.[4]

In the 1990 Clean Air Amendments Act, the federal government required refiners to seasonally mix oxygenates in gasoline to reduce pollution in certain regions with particularly severe air pollution. Methyl tertiary butyl ether, better known as MTBE, was the favoured additive until it was discovered that it was leaking into groundwater and creating a potential health hazard. Demand for ethanol as an oxygenate jumped in the early 2000s when several large states, including California and New York, began phasing out the use of MTBE. With the intent of reducing dependence on imported oil, the 2005 Energy Policy Act added a broader mandate for replacing gasoline with a minimum level of renewable fuels, mostly corn-based ethanol with today's technologies. The Global Subsidies Initiative estimated in Fall 2007 (before the mandate was expanded) that the total cost to consumers and taxpayers of the support for biofuels in the United States would be roughly US$10 billion per year from 2006 to 2012 (Koplow 2007).

As shown in Figure 5.4, however, the initial mandate levels in the 2005 Energy Policy Act were non-binding because the rising price of gasoline was driving demand for ethanol above prescribed levels. In 2007, in another effort to reduce dependence on imported oil, US Congress doubled the mandated level for using renewable fuels to 9 billion gallons in 2008 and 15 billion gallons by 2015. The mandate rises to 36 billion gallons by 2022, of which no more than 15 billion gallons should be from corn and the rest from "advanced" biofuels. Meeting this mandate will require the development of the technology for advanced biofuels, and automobile technology and regulatory policy will also have to evolve. Currently, EPA regulations prohibit use of ethanol blends higher than 10 percent, called E10, except in "flex" vehicles, and automobile manufacturers will not provide warranties for conventional vehicles running on blends of more than 10 percent ethanol because of concerns that higher levels could harm automobile engines.[5] Theoretically that suggests a real-world limit for ethanol of around 14 billion gallons, but if gasoline consumption were to remain at around 140 billion gallons annually, experts say the real cap would be around 11–12 billion gallons because of logistical constraints and state regulations that cap ethanol blends at below 10 percent (Rohde 2008).

Based on recent acreage and yields (and assuming that the technical and regulatory issues are resolved), the mandated level of 15 billion gallons by 2015 would require more than 40 percent of US corn production. Conservatively assuming no increase in gasoline consumption from the current level of 140 billion gallons annually, and taking into account that a gallon of ethanol produces only about two-thirds as much energy as a gallon of gasoline, using nearly half the US corn crop for ethanol would reduce gasoline consumption by only around 7 percent. This seems a high cost for such a small step toward reduced dependence on imported oil.

Another argument for ethanol made in the midst of the 2007–8 energy crisis was that it prevented gasoline prices from rising even higher—20–40 cents per gallon higher by one estimate (Du and Hayes 2008). But to the degree that reduced consumption must be a part of any strategy for reducing dependence on fossil fuels and thereby reducing greenhouse gas emissions, lower gasoline prices create perverse incentives.

Ethanol production capacity was just over 8 billion gallons in early 2008, with another 5 billion gallons of capacity under construction. Bruce Babcock of Iowa State University's Center for Agricultural and Rural Development (CARD) estimates that removing the mandate would, in the short run (with gasoline prices over US$3 per gallon), reduce ethanol production by 4 percent and corn prices by only 1.2 percent. Last year, the tariff and the tax credit for blending ethanol in gasoline were more important factors because they helped to offset the rising costs of corn feedstock, which squeezed producer margins.

Eliminating all three would reduce ethanol production by 21 percent but the corn price by only 12.5 percent. In the longer run, the CARD model underscores the close links between ethanol production, corn prices, and the price of gasoline. Even with the elimination of federal biofuel policies (many states have their own), Babcock estimates that wholesale gasoline prices of over US$3 per gallon would stimulate ethanol production of around 14 billion gallons, nearly the mandated level for 2015, which would keep the corn price at US$4 per bushel, roughly the average price for 2007 (Babcock 2008).

Illusory Benefits of Food-Based Biofuels

Recognizing the potential negative effects for food markets, the US Congress limited corn-based ethanol to 15 billion gallons when it set the mandate for 36 billion gallons of biofuels by 2022 to replace gasoline and home heating oil. As noted, that is similar to the regulatory limit for blending ethanol in gasoline (10 percent) given gasoline consumption of roughly 140 billion gallons annually. As noted earlier, when the lower energy value of ethanol is factored in, the mandated level would consume nearly half the US corn crop and reduce gasoline consumption by less than 10 percent. Measures to discourage consumption and promote efficiency in energy use would also reduce dependence on foreign oil without the perverse incentive of lower gasoline prices, an incentive that is at odds with sound climate change policy.

New scientific research also suggests that the climate change benefits of corn ethanol are not just small but may even be negative. Previous life-cycle analyses of the impact on greenhouse gas emissions (taking into account the energy used in producing it) suggest that corn-based ethanol can reduce emissions by roughly 20 percent, depending on the process and the fuel used to refine it (biomass, natural gas, or coal).[6] Corn is a relatively energy-intensive crop and requires large amounts of water as well; run-off from the chemicals used to grow it also contributes to water pollution and, ultimately, given the heavy production along the Mississippi, the large and growing dead zone in the Gulf of Mexico. These environmental costs are rarely calculated in assessing the alleged benefits of ethanol.

In addition, new research recognizes that increased demand for biofuels is likely to lead to new land being brought into production, either directly, to produce the feedstock, or indirectly, by bidding up food prices and encouraging increased production of food on new land elsewhere. Chopping down forests or plowing up grassland releases the carbon stored in both the plants and the soil.

One recent study calculates the "carbon debt" created when forests or native grasslands are converted to biofuel feedstock production. Depending on the type of land converted and the type of biofuel produced, the time it would take to repay the carbon debt due to land-use changes varies from 0 for grasses grown on marginal cropland, to 423 years for palm biodiesel produced from peatland rainforest. It would take more than three hundred years to repay the carbon debt from deforestation of the Amazon to produce soybean biodiesel (Fargione et al. 2008). Sugarcane ethanol is the most efficient of the biofuels examined, but even in that case, if demand for sugar for ethanol leads to land-use changes, it would take seventeen years to repay the carbon debt created (ibid.) (see also Box 5.1). The carbon debt created by corn ethanol produced from native grasslands in the United States would take 93 years to repay and 48 years were the corn grown on abandoned cropland, such as the acreage that might be released from the US Conservation Reserve Program (CRP). In addition to the carbon emissions, renewed production on CRP land could contribute to local water pollution, soil erosion, and loss of wildlife habitat.

A second study tries to estimate the impact on greenhouse gas emissions when new land is converted to grow food to replace that displaced by biofuel production. In assessing the potential impact of diverting enough corn to produce 15 billion gallons of ethanol, the authors must make a number of assumptions about the type of land and number of additional acres that might be converted for food production, as well as the location of the land, and this creates significant uncertainty regarding the precision of the estimates. Qualitatively, the results are the same as those above, finding that the net effect of increased production of corn-based ethanol would increase greenhouse gas emissions. Specifically, this study concludes that, "over a 30-year period, counting land-use change, GHG [greenhouse gas] emissions from corn ethanol nearly double those from gasoline for each km [kilometer] driven" (Searchinger et al. 2008, 1239).

Box 5.1 *Is Sugar Ethanol Different?*

Ethanol made from sugar cane is far more efficient, both economically and environmentally, than that made from corn, though the industry in Brazil, like the one in the United States, required substantial subsidies to get it off the ground. The government provided low-cost loans to processors and encouraged the development of "flex-fuel" cars to further encourage ethanol use, and it continues to require that all gasoline contain at least 20–25 percent ethanol (Goldemberg 2008). With government support to cover the fixed costs of building production capacity and the distribution infrastructure, and oil prices over US$40 a barrel, sugar ethanol is competitive without subsidies, and it replaces roughly 40 percent of the gasoline that would otherwise be consumed in Brazil (ibid., 2).

Box 5.1 *Continued*

According to one study, US corn-based ethanol costs nearly three times as much to produce as Brazilian sugar ethanol and is still 70 percent more costly after accounting for the sale of byproducts and government subsidies (ibid.). With corn, there is an extra step in production, because the starch must be converted to sugar, which is then distilled into alcohol. In addition, bagasse, the fiber left after the sugar-containing juice is extracted from the cane, is used to power ethanol processing plants, which both lowers costs and provides environmental benefits relative to the coal or natural gas most often used to process corn ethanol.

Studies also suggest that sugar ethanol contains eight to ten times as much energy as goes into producing it and that it reduces greenhouse gas emissions by around 80 percent relative to gasoline, excluding land-use changes (ibid.; see also Preto 2008). While concerns have been raised that increased sugar production for ethanol could contribute to further deforestation in the Amazon ecosystem, analyses do not suggest that is likely at least for the short-to-medium run. Sugar itself is not grown around the Amazon because it is too wet there, but ethanol production might indirectly contribute to deforestation if it displaced soybeans or cattle-grazing that was then relocated to the Amazon. Sugar cultivation currently occurs on just 2 percent of the land used for agriculture and grazing in Brazil, and studies suggest that planned expansion of sugarcane will come mostly from degraded grazing land (Goldemberg 2008). Still, there should be safeguards to ensure that ethanol expansion does not indirectly contribute to further degradation of the Amazon.

Conclusions and Recommendations

Although it is difficult to know the precise contribution of biofuels to rising food prices around the world, policies promoting production of the current generation of biofuels are not achieving their stated objectives of increased energy independence or reduced greenhouse gas emissions. Corn ethanol cannot make a significant dent in petroleum consumption without changes in automobile technology and the diversion of huge increases in the share of food (corn, soy, and palm oil) production to fuel. Nor are these biofuels contributing to slowing climate change, because new lands must be plowed to grow food. In sum, increasing hunger due to higher-cost food adds urgency to the need to change biofuel policies, but it does not change the basic fact that there is little justification for the current set of policies.

Specifically, if all of the US-mandated 15 billion gallons of renewable fuels for transportation (and home heating oil) by 2015 were blended into gasoline, it would replace just 7 percent of current gasoline consumption and use roughly 40 percent of the corn crop (based on recent production levels). Moreover, while it has long been known that the net energy and greenhouse gas emission benefits of corn-based ethanol are relatively small because its production is energy-intensive, recent scientific studies suggest that the current generation of biofuels, including biodiesel made from palm oil, soybeans, and rapeseed, as well as corn-based ethanol, actually add to greenhouse gas emissions relative to petroleum-based fuels when land-use changes are taken into account. That is, greenhouse gases are released when forests are cut down or grasslands cleared to plant biofuels, or food is planted on new acreage to replace crops diverted to fuel elsewhere. Sugar is far more efficient as a source of ethanol and may have a role to play, but the situation must still be monitored carefully to ensure that soybean and cattle production displaced from sugar-growing areas does not lead to accelerated clearing of tropical forests in the Amazon region.

At the same time, even if biofuel subsidies in the United States and the EU were reduced or eliminated, sustained attention and additional steps would still be needed to address food security, especially in developing countries. When oil prices again rise above US$60–80 per barrel, as they will, demand for ethanol as an alternative will also revive, even without government intervention. That underscores the need for conservation measures, as proposed by UN Secretary General Ban Ki-moon, World Bank President Robert Zoellick, and others (see Lustig 2008), to reduce energy use and for significant increases in investments in agriculture in developing countries.

Notes

This chapter is based on CGD Working Paper 151 (Washington: Center for Global Development), which was made possible in part by financial support from the William and Flora Hewlett Foundation and the John D. and Catherine T. MacArthur Foundation. The views expressed here do not necessarily reflect those of the staff or boards of directors of the Center for Global Development.

1 The role of financial speculation in recent commodity price increases remains highly disputed. For a detailed analysis concluding that such speculation has not played a major role, see Sanders, Irwin and Merrin 2008.

2 The ratio ranges from nearly 3:1 for poultry to around 7:1 for pork and beef (see Trostle 2008, 12). While China and India have attracted significant attention in this context, Abbott et al. 2008, point out that both countries pursue policies of self sufficiency and trade very little, thus the largest impact from increasing demand is yet to come.

3 Abbott et al. (2008) argue that the role of dollar depreciation has been under appreciated in most analyses of the food price story.

4 In fact, the United States scheduled the duty under the heading of "other duties and charges," as allowed by the Uruguay Round Agreement on Agriculture concluded in 1993. These other

duties and charges are not a subject of the ongoing Doha Round negotiations on tariffs and will not be cut if an agreement on market access is reached.

5 Flex vehicles can run on any combination of gasoline and ethanol. Brazil requires that all gasoline contain 20–25 percent ethanol, and American supporters of ethanol, including Senator John Thune (R-SD) have appealed to the administration to raise the cap for conventional vehicles, but no action had been taken at the time of writing (Green Car Congress 2007).

6 The Congressional Research Service reviewed the literature on corn-based ethanol and found that the central estimate was that a gallon of ethanol contains, on average, 20 percent more energy than the energy that goes into producing it and that it reduces greenhouse gas emissions by 10–20 percent relative to gasoline (see Yacobucci and Schnepf 2007, 9, 12; see also EPA 2007; Searchinger et al. 2008, 1239).

Works Cited

Abbott, Philip C., Christopher Hurt, and Wallace E. Tyner (2008). "What's Driving Food Prices?" Oak Brook: Farm Foundation. July.

Babcock, Bruce (2008). "Statement before the US Senate Committee on Homeland Security and Government Affairs." Hearing on Fuel Subsidies and Impact on Food Prices. Washington DC. 7 May.

Boonekamp, Loek (2008). "The Grain of Truth: Food Prices Have Risen Sharply. Why?" *OECD Observer* 267 (May–June): 16–17.

Cline, William R. (2007). "Global Warming and Agriculture: Impact Estimates by Country." Washington, DC: Center for Global Development and Peterson Institute for International Economics.

Collins, Keith (2008). "The Role of Bio-fuels and Other Factors in Increasing Farm and Food Prices: A Review of Recent Developments with a Focus on Feed Grain Markets and Market Prospects." Supporting material for a review conducted by Kraft Foods Global, Inc. 19 June.

Du, Xiaodong, and Dermot J. Hayes (2008). "The Impact of Ethanol Production on US and Regional Gasoline Prices and on the Profitability of the US Oil Refinery Industry." Working Paper 08-WP 467. Ames, Iowa: Center for Agricultural and Rural Development, Iowa State University.

Elliott, Kimberly Ann (2006). "Delivering on Doha: Farm Trade and the Poor." Washington, DC: Center for Global Development and the Peterson Institute for International Economics.

Fargione, Joseph, Jason Hill, David Tilman, Stephen Polasky, and Peter Hawthorne (2008). "Land Clearing and the Biofuel Carbon Debt." *Science* 319 (February): 1235–38.

Green Car Congress (2007). "Thune Pushes for E20 Approval by US EPA." Green Car Congress. 12 March.

Goldemberg, Jose (2008). "The Brazilian Biofuels Industry." *Biotechnology for Biofuels* 1, no. 6.

Koplow, Doug (2007). "Biofuels—At What Cost? Government Support for Ethanol and Biodiesel in the United States: 2007 Update." Geneva: Global Subsidies Initiative.

Lustig, Nora (2008). "Thought For Food: The Challenges of Coping with Soaring Food Prices." CGD Working Paper 155. Washington, DC: Center for Global Development.

Mitchell, Donald (2008). "A Note on Rising Food Prices." Policy Research Working Paper no. WPS 4682. Washington, DC: World Bank.

Preto, Ribeirão (2008). "Biofuels in Brazil: Lean, Green and Not Mean." *The Economist*. 28 June.

Rohde, Peter (2008). " New Challenges for Ethanol." Kiplinger Business Resource Center. 21 February.

Rosegrant, Mark W. (2008). "Biofuels and Grain Prices: Impacts and Policy Responses." Washington, DC: International Food Policy Research Institute.

Sanders, Dwight R., Scott H. Irwin, and Robert P. Merrin (2008). "The Adequacy of Speculation in Agricultural Futures Markets: Too Much of a Good Thing?" Marketing and Outlook Research Report 2008-02. Urbana-Champaign: Department of Agricultural and Consumer Economics, University of Illinois at Urbana-Champaign.

Schneph, Randy (2008). "Higher Agricultural Commodity Prices: What are the Issues?" CRS Report for Congress, no. RL34474. Washington, DC: Congressional Research Service. 29 May, updated.

Searchinger, Timothy, Ralph Heimlich, R.A. Houghton, Fengxia Dong, Amani Elobeid, Jacinto Fabiosa, Simla Tokgoz, Dermot Hayes, and Tun-Hsiang Yu (2008). "Use of US Croplands for Biofuels Increases Greenhouse Gases through Emissions from Land-Use Change." *Science* 319 (February): 1238–40.

Slayton, Tom and Peter Timmer (2008). "Japan, China and Thailand Can Solve the Rice Crisis—But US Leadership Is Needed." Washington, DC: Center for Global Development.

Trostle, Ronald (2008). "Global Agricultural Supply and Demand: Factors Contributing to the Recent Increase in Food Commodity Prices." Washington, DC: United States Department of Agriculture.

United States Department of Agriculture (2008). "Grain: World Markets and Trade. Foreign Agricultural Service." Circular Series FG 06-08. Washington, DC.

United States Environmental Protection Agency (2008). "EPA Keeps Biofuels Levels in Place after Considering Texas' Request." 7 August.

World Bank (2009). Commodity Price Data (Pink Sheets). Washington, DC. http://econ.worldbank.org/WBSITE/EXTERNAL/EXTDEC/EXTDECPROSPECTS /0,,contentMDK:21148472~menuPK:556802~pagePK:64165401~piPK:64165026~ theSitePK:476883,00.html.

Yacobucci, Brent D. and Randy Schnepf (2007). "Selected Issues Related to an Expansion of the Renewable Fuel Standard (RFS)." CRS Report no. RL34265. Washington, DC: Congressional Research Service.

PART 2

Immediate Governance Challenges and Proposals: Food Aid, Trade Measures, and International Grain Reserves

CHAPTER 6

Responding to the 2008 "Food Crisis"

Lessons from the Evolution of the
Food Aid Regime

Raymond F. Hopkins

The spike in early 2008 of world food prices, particularly grains, spawned widespread discussion of a new global food crisis. Although most international food prices have fallen since mid-2008, prices in developing countries have declined little. High food prices can spawn widespread negative effects, including rising illegal movements of people, higher mortality, and slower economic growth. Globalization causes rapid and broad spread of these effects but has also expanded the ability of others to recognize and respond to emergencies.

By the end of 2008, however, the food crisis was displaced by global economic crises, and international policy attention to the food crisis has been diverted by other crises—credit freezes, slowing growth, and sharp declines in wealth held in real estate and equities. A food crisis of 2009 arguably still exists, however. It is driven by high prices in some markets (southern Africa, Bangladesh) and by declining "entitlements" to food (Sen 1983). Unemployment in the cities and lower farm incomes are expected, especially in poor countries (FAO 2009). In 2008, high world prices were a problem despite promising larger returns to farmers, but that is no longer a factor. This coming farm sector depression—accompanied by increased hunger among the most vulnerable groups, including those newly unemployed—is unlikely to motivate a politically salient group. Loss of access to food receives high-level policy attention only when hunger riots break out in urban areas. The Food and Agriculture Organization of the United Nations (FAO) price trends reported in earlier chapters—and updated by a report in 2009—forecast increased numbers of people facing food insecurity in 2009 and beyond.

Although international food prices have eased, hunger threats continue. According to the FAO,[1] this is especially so for those needing emergency and humanitarian assistance. Food aid, the principal short-term solution to this crisis, has not been adjusted, as it was in earlier crises in the 1980s and 1990s. The key question to consider, therefore, is whether responses to the broad economic crises divert attention and resources away from strategies to address food crises. In the spring of 2008 the World Bank tabled proposals aimed at assisting more efficient global food production, providing greater poverty alleviation, and assuring more stability to internationally funded feeding programs. These have been echoed in proposals by IFPRI (von Braun, Lin, and Torero 2009) for stabilizing food access.[2] The April 2009 London meeting of the G20 focused primarily on the broad financial and economic depression crises, so that these and other proposals were pushed aside by global leaders.[3] So, how can food insecurity and the management of global food safety nets using food aid be included in discussions of larger problems?

A way to avoid marginalization of the food issue is to situate the food crisis as part of the larger global economic and security threats demanding attention. This entails focusing on unreliable food aid transfers, erratic food-price signals, and the inefficient consequences of erratic food prices and failed food guarantees. It further entails focusing food reforms on new regulations and instruments consistent with those coming under consideration for dealing with the general economic malaise. Measures could, for example, focus on greater management of risks, more transparency, stronger paths for smoothing markets, and hedges to prevent shocks in one sector from amplifying problems elsewhere.

This chapter draws on the history of food aid to offer some lessons for a reform strategy for food aid. This reform should be linked to reform efforts for larger economic financial crises currently plaguing world leaders. Reforms of food aid and its current central humanitarian assistance principle should link their outcomes to reducing current world threats of economic depression and the ever wider spread of terrorism.

Background

Stories about rising hunger, the marginalized poor, and political discontent appeared in the media of various countries in the spring and summer of 2008. Some focused on a global threat, others on national dangers and political discontent. These challenges are especially burdensome for people in countries requiring suddenly expensive imports, for people already depending on inexpensive staple foods, and for victims of man-made and natural disasters living in food insecure areas (Collier 2008). Such people faced new threats to their minimum dietary needs. Causes for this crisis ranged from a growing demand in China and India to wasteful use of corn for biofuel. As discussed in earlier chapters, these causes often reinforced each other; assigning blame has proved a pointless exercise for the most part. The overarching point is that access to food seemed jeopardized and the international insurance to mitigate this threat, commodity food aid, was dwindling. As usual, food aid flows declined just as high prices increased the need for it.

The early 1960s and the late 1990s saw high levels of food aid flows (over 15 million tons),[4] but levels have dwindled in spite of the 2002 Monterey commitments to increase Official Development Assistance (ODA). From 2007 to 2009 food aid was less than half of its 2002 level. Indeed, deliveries of food aid declined after 2003, even while overall levels of ODA climbed. By 2007–08 both were in decline, with food aid averaging 6.5 million tons (see Figure 6.1).

The tight world food market and sharp prices rises for all cereals in early 2008 underlined the deep interdependence among world food markets. Markets and prices are linked far more than in the previous crises of the 1970s and 1990s. With local markets reacting quickly to changes in import prices for key foodstuffs, reactions of anger and panic emerged in dozens of countries. Thus the substantial upward shift of demand between 2000 and 2006, along with slower growth of production, translated quickly into rising commodity prices.

Figure 6.1
Food Aid Flows, 1999–2007

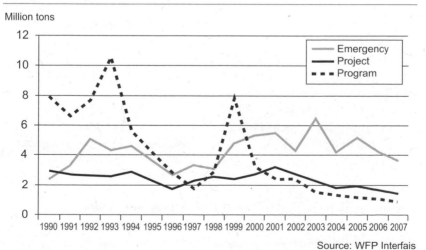

Source: WFP Interfais

This led to alarm that the poor would face new burdens, that governments would be held accountable for the sudden price rise, and that victims of natural or man-made disasters would suffer from inadequate emergency food transfers through the World Food Programme (WFP) or private nongovernmental organizations (NGOs) whose ability to secure food was undercut by high prices.

Paths for innovation in today's food aid regime must reflect and build on these larger external political and economic forces. This a central lesson of past regime reform. Although the national political forces and international crises that shaped changes in food aid over the past sixty years are substantially different today, the links between high politics and outcomes in sector-specific and welfare-related global policy continue. Thus a reform path today must link changes of food aid practices and regulations to a reduction of global financial, economic, and political instabilities. Managing public resources for food aid — now almost exclusively for humanitarian ends—requires modifying the current practice of making appeals and delivering donated food (OCHA 2008). Although the WFP functions well among UN bodies, and its expansion of donors and coverage since 1992 is impressive, it would benefit by adopting an insurance approach for its work, a change in rules rather than principles of the regime. This would allow counter-cyclical performance of emergency aid, thus reducing instability in the regime, and create incentives for those insured to better address their own national production and organizational resources to assure food security.

Moving toward use of insurance principles relates well to the larger policy forces at work—ones scrambling to reshape global economic rules that allowed financial market meltdowns globally. Insurance-governed flows of food aid that succeed in stabilizing resource flows and assure that needs are met more equitably would parallel the type of changes sought to address the unreliability discovered in our broader economic system. Such reforms could be part of the larger package of goals and reforms likely to emerge from global negotiations over finance and reserves that seek to assure that credit and trade flow efficiently. Other schemes to stabilize prices have been recommended, ones requiring coordinated stock holding and use of futures markets (von Braun, Lin, and Torero 2009). Such practices are compatible with an insurance approach. Indeed, they might be methods for achieving the risk-sharing and counter-cyclical features that are necessary components of effective food aid reform.

Three Eras of Governance in the Food Aid Regime

The principles, norms, rules, and procedures for food aid have undergone numerous changes since World War II, some sweeping, some consolidating, and others incremental. For the most part, these changes have paralleled ones in other major regimes, notably those governing global trade, banking, finance, and exchange rates. The major elements in the evolution of food aid, as shaped by these three regime eras—a surplus disposal regime, a developmental regime, and the current humanitarian assistance regime—reveal lessons for when changes occur (during crises and following policy changes in related areas) and how change events are complex. Each regime contained elements of the others but each also relied on political support from specific and different dominant forces in international politics—ones essentially outside the food sector (Ruttan 1993).

Surplus Disposal

The surplus disposal regime emerged after World War II from a mixture of domestic and international political motivations; it was driven by the hegemonic power and interests of the United States. The same conditions also drove the sweeping changes that launched the United Nations, the international financial institutions (IFIs), and the Marshall Plan. Large portions of food trade after World War II consisted of assistance flows to help reconstruct former war-torn countries, including defeated opponents of the US. With recovery underway by the early 1950s, US farm policy creating growing stocks of government-owned grains, aid outlets for production shortfalls in Asia (Korea, Pakistan, and India) proving to be sporadic, and with diplomatic tools in demand to use in the "war on communism," the US enacted a permanent program for food aid (PL480) in the summer of 1954. Its goals included assuring farmers that government stocks would not be released domestically and indeed might be used to open markets overseas once recovering countries could afford commercial imports (as occurred in the United Kingdom, Korea, and elsewhere). This 1954 US creation was complemented by the establishment of the Consultative Subcommittee on Surplus Disposal (CSSD) of the FAO, a reactive international policy to prevent blatant surplus dumping to advantage US farmers (still in place today under its Commodity Problems Committee). The CSSD set the principal bounds for a food aid regime in which the US gave 10–15 percent of its foreign aid as tied food linked to justifications such as military base rights or humanitarian needs.

Food aid in this period was a type of subterranean trade regime. The food traded as aid was linked to broader rules for trade in food commodities. The regime expressed broad nation-state agreements as well as specific national actions. The regime was and is partially separate from normal food trade, and unique organizations and rules were adopted to exempt and regulate food aid flows within broader arrangements—the International Wheat Agreement, the special rules for agriculture under the General Agreement on Tariffs and Trade (GATT), and the exemptions for food aid for below-market exchanges that could avoid being pure dumping or market-capturing strategies. At this point up to 30 percent of some commodities entered international trade as food aid. Canada joined other exporters, such as Argentina and Australia, in watchful suspicion as food aid was scrutinized under the CSSD to be sure that countries getting aid (except for dire emergencies) also purchased amounts equal to their usual imports. This was to guarantee that aid would act as a counter-cyclical force to domestic and international swings and be "real"; that is, not be just an unfair trade practice parading as a gift or subsidized loan to a poor state. While the US was the principal target of such scrutiny, other exporters—as they became donors—were also subject to these rules and to occasional complaints.

More donor countries were added, and many Organization for Economic Cooperation and Development (OECD) states made commitments under the Food Aid Convention (FAC), an agreement produced as the last agriculture bargain of the Kennedy GATT round in 1967 which called for a food aid "burden-sharing" arrangement among richer states. Even as this regime was consolidating, a newer set of regime principles was emerging. These rule developments led to a regime shaped by new events. The food crisis of 1974 brought pressure to focus food aid away from its previous role as a political and market development tool. The 1974 World Food Conference saw the World Bank and others champion poverty-reduction goals for aid, which prompted a shift in the rationale for food aid toward principles centred on development.

Development

The rise in the mid-1970s of economic development as the core principal in the food aid regime emanated from rhetorical shifts of leaders, in demands for measures of food aid's impact on development, and the creation of new, or the adaptation of existing, organizations focused on development. These included new charges given to the WFP, enhanced lobbying by international NGOs, the creation of high-level UN bodies such as the World Food Council, and lobbies for food for development in a number of countries, such as Bread for the World in the US.

In this manner, the development principle holding that aid to poor countries should be targeted for development was elevated, especially so if it were food. This also led to the priority of tracing amounts supplied to the poorest countries and to the most food insecure populations in such jurisdictions. As surplus disposal motives weakened following major domestic farm policy changes in the US and later the EU, even more important changes in world affairs enabled the regime shift to occur.

Promoters of a multilateral, development-centered use of food aid had earlier led the way to the creation of the WFP in 1963. At this time they began to promote a focus on protection and development for poor countries in the UN forums, including the United Nations Development Programme (UNDP) and the World Bank. These new or amended organizations served as agents for the development principle's elevation in the food aid regime even before it emerged as a dominant theme for food aid in the 1970s.

Equally central to the ascendance of development as the overarching principle of food aid in the 1970s were major global political and economic changes: the emergence of Détente (the ease of tensions from the Cold War), price shocks from skyrocketing oil and food prices in 1973–74, the 1971 collapse of the US dollar as a convertible currency (to gold), the end of the Vietnam war, and the assertion of Europe and Japan as powerful important allies of the United States. Suddenly, economic stability and defense were more multilateral, more interrelated, and more demanding of accommodations among North–South countries. These shifts in global politics made it possible for seeds of development principles to ascend to the top of the food aid regime.

The core ideas of this "new" regime can be found in the documents of the World Food Conference of 1974, in the changing laws of the US congress, and the OECD's Development Assistance Committee (DAC) rules regarding eligible recipients of aid. It was based on the ideology that economic development was not only possible for poor countries but was also largely a matter of getting correct policies and adequate funding in public flows. Development as the central goal of food aid was a dominant orientation among food aid professionals by the 1980s (Hopkins 1992).

Examples of successful development within countries that had been significant food aid recipients added empirical verisimilitude to the arguments of theorists and moralists. Korea, Taiwan, Brazil, Indonesia, Jamaica, and even Israel emerged as celebrated cases where food aid had been a positive force for development. Food aid shifted from Asia to Africa.

Within the development-oriented regime, however, food aid maintained some of the original motivations that launched it. It continued to be used for diplomatic ends, its allocation was still shaped by Cold War and anti-communist sentiments, and the interests of farm exporters continued to shape the total amounts and regime rules (such as where most of the food would be purchased). These forces did not disappear, they just became decreasingly salient.

In addition, other concerns about distortive and disincentive effects of tied aid also troubled many development enthusiasts who frequently condemned food aid as a problematic and possibly counterproductive resource for development. Such criticism weakened the appeal and strength of the development principle and laid the ground for an easy shift to reducing the portfolio of "development" or "project" food aid—most of which was tied to commodities grown in donor countries—in later decades. This viewpoint emphasizing bad effects from food aid was pointed out in a seminal article by Ted Schultz in 1960 and was debated among academics and officials in aid organizations for years (see Abdulai, Barrett, and Hoddinott 2005).

Humanitarian Assistance

The collapse of the Cold War and rise of large-scale emergencies needing food aid, especially in Africa, shifted the focus of food aid and its central rationale to emergency and humanitarian relief by the 1990s. A stable and durable consensus had emerged by this time around the collective international goal of eradicating hunger. This principle has come to supersede other principles governing the behaviour of states regarding food aid and trade issues, development, and political goals, and it explains the expansion of support for international food aid and hunger reduction.

While humanitarian concerns were always a component in regime norms, and allocations shifted as needs arose, actual negative correlations can be found between the domestic production of recipients and food aid received for large stretches between the 1950s and 1970s (Hopkins 1984). Examples include Iran, Israel, Egypt, and other important foreign policy recipients. Even in the 1980s, when development concerns were enhanced, governments of countries regularly flaunting their own populations' needs received food aid—as occurred in Zaire and Ethiopia, for example. By the 1990s new modalities using NGOs and forcing governments to monitor use of aid emphasized that the top priority for food was to alleviate emergency needs. Such rules do not solve problems such as those posed by Zimbabwe's refusal to accept and dispense food aid in 2008, or similar withholding of aid by a government such as Myanmar or North Korea, but they do provide leverage for negotiating better results for those threatened by hunger.

The major problem with food aid today is similar to that which occurred in 1973–75—as need grew, supply dwindled. Such pro-cyclical availability is a function of rising prices but only in part. Suddenly needy countries and peoples— shocked by natural disasters or war—must seek aid by demonstrating that people are dying, and donors make pledges yet need not ever honour them. This is a defective framework for international transfers and stabilization against hunger threats. This flaw in the humanitarian regime exists because the of earlier idea that donor control was a virtue, the overarching goal being that recipients were influenced by aid—whether by adopting "correct" development policies or strengthening political alliances and relieving the overhang of the donor government's stocks.

Because the World Food Programme handles over half of the world's food aid today, and its portfolio of projects has shifted from 25 percent to 75 percent emergency aid since the 1990s, one vision would be that it have autonomy like the World Bank to raise capital broadly or through replenishments that gave it considerable latitude. This has never been true for the WFP or for food aid generally. Replenishment pledges acknowledged under the Food Aid Convention have declined. Emergency appeals have grown to the point of becoming permanent institutionalized means for some recipients' food aid. Experiments to use insurance, the way farmers do for their crops, or national health schemes, have been tried in Ethiopia related to rainfall and production and in Malawi related to national import needs and prices. Although both were evaluated as successful, neither has continued, let alone spread. Insurance for poor countries, even if subsidized as it was in these cases, seems much like the old adage about why people do not fix leaking roofs—it's not necessary when it's not raining and not plausible when it is.

Lessons for Today

Shortcomings in the humanitarian assistance regime's performance, accompanied by a rise in demand for quicker, more reliable response to shocks such as those caused by the 2008 price spike, invoke lessons regarding the occurrence of regime change. Regime change has come to food aid about every twenty years, and since the humanitarian dominant principle ascended in the early 1990s, time for another change seems imminent. A successor to the humanitarian regime will reduce risk and insure national safety nets, much as the International Monetary Fund (IMF) now is predicated to do for international finance.

Precursors to an insurance-oriented regime can be found as early as the creation of an IMF food facility in 1981. New thinking, emerging from high-level international meetings,[5] lays a foundation for reform of food-related humanitarian aid that moves beyond the present ad hoc system of appeals, often delayed or inadequate responses, and pro-cyclical resource availability. For example, the declaration of the 2002 World Food Summit calls for efforts "to improve the effectiveness of emergency actions" including expansion of "the scope and coverage of social protection mechanisms, in particular of safety nets for vulnerable and food insecure households," in order to eliminate the threat of famine for good (FAO 2002). The 2005 Ministerial Meeting of the World Trade Organization (WTO), held in Hong Kong in December, agreed to bring food aid that serves as a disguised export subsidy under "effective disciplines" (to be agreed upon in the future), but also created a "safe box" to assure that no impediments are imposed vis-à-vis "bona fide" food aid, especially emergency assistance (WTO 2005). Donor agencies have recently moved to reform the ways in which they deliver food aid, with Canada authorizing the purchase of up to 50 percent of its assistance in developing countries (CIDA 2005). US policy-makers are discussing how to make cash available from the food aid budget for local purchases in recipient countries or in neighbouring developing countries in the case of emergencies; efforts to gain legislative authority for such trans-fers failed in 2005.

Other established practices for food aid lay a foundation for this regime change. These include rules to stabilize food flows and make them more effi-cient, such as triangular transactions, emergency release of food aid stocks aimed for development, and the substitution of cash and local purchases by several donors and the WFP. These measures all focus on quickly and effectively getting food to those suffering the shock of an unexpected accidental fall in availabilities. Efforts to end or reduce US tying of food aid to US-grown com-modities are another sign of discontent and demand for change consistent with delivery of compensatory relief.[6]

Resources to fund an "insurance" arrangement regime may also be on the horizon. Consider that the pledge from an FAO Rome Food Summit in June 2008 called for several billions of dollars in pledges mainly for agricultural development. Amounts multiple times more were made available in the follow-ing year, with over US$1 trillion in stabilization and stimulus funds for devel-oping countries pledged at the April 2009 G20 meeting. The world has witnessed trillions of dollars lost in financial markets, forcing governments to spend even more trillions on propping up private banks, insurance firms, and equity/lending markets, so it seems consistent to commit to reform for similar stabilization and restoration for these countries' food systems and food safety nets.

Currently aid that arrives in poor countries following shocks to the food system is often late, poorly targeted, and inadequate. These shortcomings are worse where national safety nets have been neglected or marginalized by lead-ers who rely instead on external aid (Sen 1983). In these cases aid has not served to reinforce national safety nets nor has it made recipient countries more shockproof or their citizens' risks more manageable. Yet such safety nets serve national economic and political objectives extremely well (see Adato, Ahmed, and Lund 2004 and Alderman and Haque 2005).

Organizing humanitarian aid guided by insurance-like principles addresses these problems. First, such a reformed regime can transform the relationship between recipient and donor governments from a particularistic one of suppli-cant and providers of charity, to a more rational–legal relationship of policy holder and insurer or re-insurer. Humanitarian assistance based on insurance prin-ciples makes eligibility for aid the right of a policy-holder in the country affected by shock. Second, the aid flow mechanisms become less tied to individual donors' immediate circumstances, such as ease of securing emergency legislative appropriations, ability to divert existing aid resources, or the existence of surplus food stocks (Gilligan and Hoddinott 2006). Third, to the extent that donors agree to this new approach to emergency food aid, it offers the opportunity to harmonize the presently disparate approaches of various humanitarian assistance agencies, thereby promising additional efficiency gains. Finally, such re-insurance arrangements increase the incentives in covered countries to develop and maintain effective national food and protective safety nets consistent with their resources.

The task of overcoming the financial crisis of 2008–09 assumed critical proportions by the end of 2008. The continuing fight against hunger, however, has not benefited from this massive increase in public-sector commitments. As argued initially, only by linking changes in food aid to broader financial security goals, reaching beyond while incorporating the classic concern of food security, will the long-identified flaws in the existing food aid regime be addressed. Making food aid an instrument of counter-cyclical global macro-economic policy is the most promising vehicle for reform. This approach, framing food aid as an insurance back-up to national safety-net and stabilization policies, would promote reform proposals as ways to reinforce existing tendencies among G20 countries to create mechanisms to ward off a deeper recession. Most proposals discussed at the April G20 summit, for example, looked to expand and harmonize regulation of key financial flows, to guard against excessive risk taking, and to spread the burden of adjustment to shocks among many counties and agencies. These are the ineluctable elements that a reform of food aid must embrace and address.

The adoption of insurance principles as part of today's humanitarian food aid regime will improve guarantees of access to food, especially when emergencies occur and food prices are high. As noted by the FAO, these changes do not require the large amount of funding demanded for large-scale changes being considered or implemented outside the food area. These external forces for regime change create openings for inserting new ideas and rationale into the practices and rules of food aid, ones that accord with solutions to these larger problems, while advancing the integrity of the regime's own principles and norms. A virtual reserve, as proposed by the IFPRI in March 2009 (von Braun, Lin, and Torero 2009), is consistent with the principles by which insurance operates to build assets available for release to those harmed. It also emerges in consistency with demands to fix the way markets operate and to avoid risk by insurers that characterized financial agencies in the last decade.

Links to these other concerns are natural. Consider that many problems in today's international environment include not only the political instabilities that breed terrorism, piracy, and a series of global "bads" (Naím 2003) but also the economic vulnerabilities across borders that exacerbate erratic behaviour in markets and financial "meltdowns." Recall that the Bretton Woods organizations of the World Bank and the IMF, and the GATT (now WTO), were created to assist national governments to share risks and regulate cross border transactions in the wake of war and depression. Security and peace-keeping aims were assigned to the new United Nations, along with other special agencies that grew over the years, from the United Nations Children's Fund (UNICEF) to functional agencies in education, food, health, and so forth. This was a unique and fertile era of regime formation and extension of rules of national restraint in order to secure more important global public goods. The end of the Cold War failed to achieve reform and regime construction of any comparable magnitude. But it did provide an opening for important shifts in regimes—nuclear weapons control, the role of the North Atlantic Treaty Organization (NATO), and many incremental changes that strengthened supra-national bodies like the European Union and routines for dealing with regional economic crises like those in Asia in 1997.

The current global financial crisis will resolve itself at some point. The April 2009 G20 meeting, however, left much undecided regarding enhanced coordination (harmonization) of regulations and the actual additional resources pledged for development of less developed countries. Much of the US$1 trillion is not new, and other funds are generated by the IMF. The resolution of financial crisis concerns, however, may not occur promptly, and not without major tragedies. Sweeping reform of international institutions and regulations is a key concern among developed-country governments. There is yet little talk of linking food security and food aid to these broader economic reforms. The path for resolving this global crisis contains a number of cooperative arrangements for burden-sharing, for ending unregulated financial instruments (such as in hedge funds, insurance of debt-equity swaps, etc.), and for counter-cyclical measures to act as circuit breakers for downward economic contagions. Fortunately for reform-minded food crisis responders, these measures are consistent with the idea of backing food security assurances with insurance principles led by an international organization, most likely the WFP.

A number of questions for research arise from this claim. Who pays for such insurance? How are claims assessed and paid? What role is there for public and private entities to share the provision of such a public good? Where would political support for such an undertaking emerge?

Studying the rise and fall of aid demands and responses is one way to track the changes in attention to hunger and humanitarian aid. The decline of contributions to calls for assistance coordinated by the UN suggests that there is a downward shift in attention given to these issues. Figure 6.2 shows results from the UN Office of Coordination of Humanitarian Affairs (OCHA) tracking appeals and responses for humanitarian aid. Key points in 2008/9 are, first, the growing shortfall of contributions compared to identified requirements, and second, the shift in requirements, so that assistance is often related to shocks from failures in water, electricity, and other systems and not the food system per se.

Conclusion

The long-term drivers of the food crisis of 2008 identified at the height of the price rises, such as growing demand from newly enriched countries, wasteful use of farm products for fuel and feed, low investment in agricultural research, and harm from the early effects of climate change, will be irrelevant to current debates unless they are linked to better risk management and the creation of counter-cycle instruments for smoothing shocks, rather than amplifying them as has happened in the 2008–09 financial crisis. To preserve and improve the global system for assuring food access for those most in need, arguments must focus on global

Figure 6.2
*UN Office for the Coordination of Humanitarian Affairs (OCHA),
Requirements and Contributions*

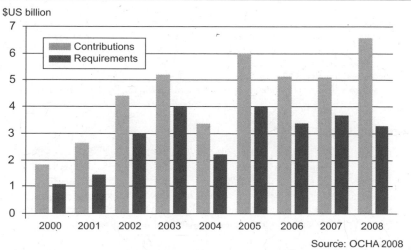

Source: OCHA 2008

macroeconomic and political stabilization. These, along with humanitarian objectives, must be the principal goals for a reformed food aid regime.

The logic of this approach begins with recognizing how erratic commodity prices have led to negative economic consequences broadly. High instability in world prices, including and particularly food prices (which account for a large fraction of poorer families' household expenditures) exacerbated these other instabilities—both as cause and effect. Instead of the underproduction feared a year ago, food production now exceeds demand. The destabilizing effects of unreliable markets and high price fluctuations fuel consumer fears, slow investment, and distort capital allocations. These factors, in which food markets play an important role, are the central motivators for drawing international attention toward the proposed stabilization of food access using insurance principles in the extant food aid regime. Addressing food and hunger crises as unique is unworkable; they must be placed within the search for international policy-makers' solutions to these larger concerns.

Notes

1   The FAO reports that more people are likely to fall below the hunger threshold following the price rise in 2008. As of March 2009, it estimates that 963 million people, or about 15 percent of the world's population, are suffering from hunger and malnutrition, a rise of 75 million people since 2007 and the director general of the FAO Jacques Diouf told the *Financial Times* it had passed 1 billion (FAO 19 March).
2   This proposal by IFPRI is further discussed by Frederic Mousseau in Chapter 8 of this volume.
3   Jeffrey Sachs, however, finds optimism in strategies discussed within USAID and by officials in transgovernmental aid networks. He sees more attention to programs for farmers and the poor in which they design the steps forward (see Sachs 2009).
4   The tonnage amounts in the 1960s are especially notable, because they accounted for up to one-sixth of world cereal trade, a far larger portion of trade than ever since.
5   The meetings include the 1996 World Food Summit and its 2002 follow-up summit, the 2000 UN Millennium Assembly, the 2002 Monterey Conference on Financing for Development, and the G8 summits of 2004 and 2005.
6   These issues are discussed more fully by Gawain Kripke in Chapter 9 of this volume.

Works Cited

Abdulai, Awudu, Christopher B. Barrett, and John Hoddinott (2005). "Does Food Aid Really Have Disincentive Effects? New Evidence from Sub-Saharan Africa." *World Development* 33, no. 10: 1689–1704.

Adato, Michele, Akhter Ahmed, and Francie Lund (2004). "Linking Safety Nets, Social Protection, and Poverty Reduction: Direction for Africa. 2020." Africa Conference Brief no. 12. Washington, DC: International Food Policy Research Institute.

Alderman, Harold, and Trina Haque. (2005). "Countercyclical Safety Nets for the Poor and Vulnerable." Paper prepared for Workshop on Food Price Risk Management in Low-income Countries, 28 February–1 March. Washington, DC: World Bank.

Canadian International Development Agency (2005). "Canada Opens Food Aid Purchases to Developing Countries." Press release, 22 September. http://www.acdi-cida.gc.ca/CIDAWEB/acdicida.nsf/prnEn/JER-32714474-R82.

Collier, Paul (2008). "The Politics of Hunger." *Foreign Affairs* (November/December): 67–79.

Food and Agriculture Organization of the United States (2002). "Declaration of the World Food Summit: Five Years Later." Rome: FAO. http://www.fao.org/DOCREP/MEETING/005/Y7106E/Y7106E09.htm#.

——— (2009). "Interactive Internet Tool Covers 55 Countries—Shows Food Prices Locally Have Yet to Fall." Rome: FAO. 19 March. http://www.fao.org/news/story/en/item/10693/icode/.

Gilligan, Daniel, and John Hoddinott (2006). "Is There Persistence in the Impact of Emergency Food Aid?" FCND Discussion Paper 209. Washington, DC: IFPRI.

Hopkins, Raymond F. (1984). "The Evolution of Food Aid," *Food Policy* 9, no.4: 345–62.

——— (1992). "Reform in the International Food Aid Regime: The Role of Consensual Knowledge." *International Organization* 46, no. 1: 225–64.

Naím, Moisés (2003). "The Five Wars of Globalization" *Foreign Policy* (Jan./Feb.): 29–37.

Office of Coordination of Humanitarian Affairs (UN) (2008). Appeals and Funding: The Financial Tracking Service (FTS). http://ocha.unog.ch/fts2/pageloader.aspx?page=emerg-emergencies&section=CE&Year=2008.

Ruttan, Vernon (1993). *Why Food Aid?* Baltimore: Johns Hopkins University Press.

Sachs, Jeffrey (2009). "Home Grown Aid." *New York Times*, 9 April, p. A23.

Sen, Amartya (1983). *Poverty and Famines: An Essay on Entitlement and Deprivation.* Oxford: Oxford University Press.

von Braun, Joachim, Justin Lin, and Maximo Torero (2009). "Eliminating Drastic Food Price Spikes – A Three Pronged Approach for Reserves." IFPRI Note for Discussion, March. Washington, DC: International Food Policy Research Institute.

World Food Programme (2008). Food Aid Flows. http://www.wfp.org/interfais/index2.htm#.

World Trade Organization (2005). "General Council's Decision on the Doha Agenda Package." Geneva: WTO.

CHAPTER 7

Preparing for an Uncertain Global Food Supply

A New Food Assistance Convention

C. Stuart Clark

Over the past decade, a growing consensus has emerged that the global human food supply is becoming increasingly vulnerable to serious disruptions. These disruptions are likely to be regional in nature but may also have global implications, as they did in the early months of 2008. Food system disruptions can have profoundly destabilizing effects on the affected countries and beyond. There is therefore a strong rationale for designing a global food "safety net" that can bridge the disruption of these food supplies.

The Food Aid Convention (FAC) is the only treaty that provides a predictable multilateral mechanism for making food transfers available in response to structural food deficits and food crises. The current version of the treaty was renegotiated almost a decade ago and in many key aspects has been rendered less and less suited to contemporary realities. Although its reform has been stalled by agricultural trade negotiations at the World Trade Organization (WTO), there is a growing consensus among the current member states that now is the time to seek significant changes, through reform of the Rules of Procedure and/or renegotiation of the convention itself. Either option has the potential to provide a global food safety net.

Discussions have already begun about the nature of any new or reformed convention, examining ideas such as:

- modifying the commitment structure to permit greater flexibility in the nature of countable contributions (e.g., non-food transfers such as vouchers and cash transfers, micronutrient supplements, cash contributions toward the costs associated with another members' commodity contributions);

- improving the effectiveness of food assistance activities by implementing a system of periodic reviews of selected interventions;
- broadening participation in the convention to include recipient countries who undertake to adhere to a code of conduct for food assistance effectiveness; and
- implementing a rights-based approach by linking the overall commitment level to a reliable measure of need and using a pro-rated formula for member contributions.

The broadening of the application of the FAC has led many proponents to promote the changing of the title to "Food Assistance Convention," underscoring the use of tools beyond food aid to increase food consumption among food-insecure people. The outcome of these negotiations could significantly improve the governance of international food security by ensuring a just allocation of food supplies for future major food crises.

Background to the Food Aid Convention

The FAC was established in 1967 in conjunction with the Kennedy Round negotiations of the General Agreement on Tariffs and Trade. Its original purpose was to provide an agreed-upon framework to regulate the use of food aid as a mechanism to utilize surplus food stock, in particular to minimize the disruption of the international trade in cereals and to provide a reliable minimum quantity of food aid. The initial membership included all the major wheat-exporting countries and some of the major cereals-importing countries. The FAC has been renegotiated five times, resulting in changes in the overall quantities of food aid, the range of "countable" commodities, and the agreed-upon purpose and qualifications for food aid transactions (see Figure 7.1).

Within certain convention periods there were adjustments in commitment levels, usually based on unilateral notifications by donors. Other changes during the past twenty-five years have included local purchase flexibility (FAC 1986) and an increase in the number of countries on the eligible recipients list, most notably the addition of the emerging economies of the former Soviet Union in the 1995 convention. The range of eligible food commodities was increased, extending beyond cereals to include limited amounts of pulses, edible oil, root crops, skim milk powder, sugar, seeds, and micronutrient fortification/supplementation (FAC 1995; FAC 1999). Eligible costs were adjusted to include those associated with transport in the case of internationally recognized emergency situations (FAC 1999).

Figure 7.1

Minimum Food Aid Tonnage Commitments under Various Food Aid Conventions

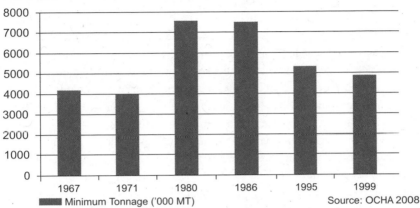

The most recent FAC was due for renegotiation in 2002 but has been extended several times pending the conclusion of Doha Round WTO negotiations on trade disciplines for food aid. Although these negotiations are not yet complete, there is sufficient closure on the food aid discussions for member states to begin discussion on FAC reform.

This renegotiation comes at a particularly important moment. Food aid needs have risen steadily over recent years, reflecting continuing low investment in agriculture and rural development, unfair agricultural trade relationships and the growing impact of climate change, which is expected to steadily reduce food production in equatorial countries. At the same time, total food aid availability has declined, with rising prices decreasing budgetary allocations for food aid. Because food is such a basic necessity, acute food shortages can quickly lead to serious political instability, particularly in urban areas. If such instability arose in several countries at the same time, the threat to the international economic and political systems would be severe. In this context, the existence of an adequate global food aid safety net may be a key element in ensuring global economic and political stability.

The Value of the Food Aid Convention

There has been considerable debate about the value of the FAC, most notably by Charlotte Benson (2000) and, more recently, by Hoddinott, Cohen, and Barrett. (2008). On the positive side it has been noted that the Food Aid Convention is the sole international treaty that guarantees a minimum transfer of resources between (mostly) Organization for Economic Co-operation and Development (OECD) countries and developing countries. By specifying a minimum quantity of food, it provides some protection from reductions in food aid availability when the food prices rise suddenly as they did in early 2008. In principle it is therefore a legal instrument for providing an international food safety net, a fact that could be of great importance in the volatile years ahead.

In practice, the FAC has fallen short of its potential. Benson points out that the total commitment level of the FAC has generally been well below the actual food aid levels, suggesting that at least some convention members set their commitment levels so low that they are unlikely to fall below them. Despite this, meeting the convention commitments has been a significant factor in some member states' allocation of resources, even though at times they threatened to weaken their commitments. The strong domestic political profile of food aid in some member states may also support the claim that these resources have been at least partially additional to other aid flows.

There are several other criticisms of the current convention. Some critiques focus on the complexities of the rules and procedures. In particular, it has a very complex and non-transparent method for rendering the quantities of a wide range of permissible commodities, delivered through several different modalities, into a single unit of food aid, the "wheat-equivalent tonne."[1] It also has complex and sometimes perverse rules regarding the accounting of transportation costs—most recently the inability to count transportation costs paid for the transport of non-member food aid shipments (e.g., Indian wheat provided for food emergencies in neighbouring countries).

Further critiques focus on what the convention fails to do. For example, it fails to fully recognize the importance of the nutritional adequacy of food aid, particularly the important role of micronutrient supplementation. There has also been a failure to implement the convention provisions that focus on ensuring the effectiveness of food aid. Although these are aspirational in nature, there is much that could be done to give them substance. The convention also fails to provide adequate representation to recipients because it limits its membership to donors only. This is a direct challenge to the emerging aid consensus represented by the Paris Declaration on Aid Effectiveness that recipients must also be represented. Finally, it lacks transparency in that reports of member compliance with their commitments are not publicly released and have been difficult or impossible to obtain.

Despite these weaknesses, most member states think that scrapping the convention would carry an unacceptably high political price—particularly in the context of an increasing need for food-related transfers due to climate change-related emergencies and instability, and volatility on world food markets. The principal debate now centres on how much to change the convention, and in what direction.

Possible New Directions for the Food Aid Convention

The debate among members about the future of the FAC has just begun. At the semi-annual meeting of the FAC in December 2008, members began informal technical discussions about changes to the convention—initially following two possible lines of reform. Some members favoured immediate renegotiation; others favoured an initial revision of the Rules of Procedure. The context of these discussions played a significant role in determining the outcome. Important elements included:

Economic conditions
Donor governments dealing with recession and falling tax revenues will be disinclined to undertake additional international commitments (and some may be tempted to drop existing commitments).

Food aid demand
The World Food Programme reports that the need for food aid is escalating as a result of both global economic conditions and the increasing impact of climate change.

WTO trade negotiations
In the context of WTO negotiations, developing countries still point to the 1994 Marrakech Decision in which rich country members of the WTO committed to several measures, including enhancing the FAC, should food prices rise as they have for the past several years.

Evolving best practices in food aid
In recent years the use of vouchers and cash transfers to households and individuals have been shown in some situations to be as effective or more effective than food transfers in responding to food insecurity. Neither vouchers nor such cash transfers can be counted against commitments under current convention rules.

Donor policies
The European Commission took a decision in 2006 to move all of its food aid activities to the European Community Humanitarian Organization (ECHO). The main mandate of ECHO is to respond to immediate emergencies, and, under current European Commission rules, it cannot make global binding commitments to provide resources on multi-year basis. ECHO can legally commit

resources on an event/project basis only. ECHO also seems to be more comfortable with the idea of transforming the FAC into a humanitarian assistance convention, thus avoiding concrete food aid commitments with their attendant price risks.

Minor Modifications—Revision of the Rules of Procedure

Several of the criticisms of the current convention could be addressed with minor changes, possibly by amending the current rules of procedure. Below are some suggestions for inclusion in these amendments:

Twinning to support food aid by developing country food aid activities
Under the current rules of procedure, it is not possible to count cash contributions that assist in the transportation and distribution of food aid commodities provided by other donors. With several developing countries expressing interest in the possibility of providing food aid resources to countries in their vicinity, several convention members, including Canada, would like to be able "twin" with these countries to cover the non-food costs.

Micronutrient supplementation
The current rules of procedure provide for the micronutrient fortification (adding micronutrients to food aid commodities) of convention-permitted commodities but not for the provision of micronutrient supplements (pills that can be separately distributed). With micronutrient malnutrition now widely documented as a major problem, there is a strong case for providing supplements either on their own or in conjunction with, for example, unfortified locally procured food aid commodities.

Increased transparency
The current rules of procedure set deadlines for member reporting of food aid activities but make no definite provision for the timely public release of FAC reporting. This could be provided for by an amendment of the current rule on reporting.

Conversion to a Food Assistance Convention

Observers of the FAC have been looking for ways to "retool" the convention to better suit current best practices. The inclusion of vouchers and cash transfers to the traditional food transfer has given rise to the call for a Food Assistance Convention. The Trans-Atlantic Food Assistance Dialogue (TAFAD), an NGO coalition, has been one of the most active voices in this debate to date. TAFAD is made up of most of the major North American and European food aid programming NGOs, and issued its first proposals for a new Food Aid Convention in August 2006. TAFAD called for several major changes to the convention in the context of meeting the first Millennium Development Goal—to halve, between 1990 and 2015, the proportion of people who suffer from hunger and the proportion of people whose income is less than US$1 per day:

1. The objective of the convention should be to meet global food aid needs based upon comprehensive needs assessments, using recognized global norms such as the Food Security Assessment and Nutrition Assessment procedures outlined in the Sphere Project handbook (Sphere Project 2004).
2. The total commitments in a future convention should be adequate to meet all legitimate food aid needs, initially estimated as equivalent to a nutritionally adequate diet for 25 million people. The retention of a commitment to an amount of food rather than a cash value is an important hedge against food aid availability being negatively affected by food price increases.
3. Food aid accounted for under the convention should be solely directed to direct food transfers (or their equivalent in vouchers or cash) to food-insecure people. This would prohibit various types of food aid monetization from being accounted for under the convention.
4. There should be timely and transparent public reporting of all food aid activities accounted for under the convention.
5. Recipient states' participation in the convention should include the obligation to adhere to a "Code of Conduct for Food Aid" and to play a role in mutual accountability for the effectiveness of food aid activities within their borders.
6. A multi-stakeholder technical advisory committee should be created, charged with overseeing the assessment of food aid needs and a peer review mechanism for food aid activities under the convention.

In May 2007, the German Government convened an international conference, "Food Aid—Exploring the Challenges," which focused its attention on the future of the convention. The conference was attended by representatives of all but one of the FAC member states, ten recipient countries, and many NGOs and research organizations. The result of the conference, dubbed the "Berlin Consensus" (Cohen and Weingärten 2007), called for the following reforms to the convention:

1. Broaden the convention to include all food assistance tools—food transfers, vouchers, and cash transfers.
2. Make the overall commitment level meaningful (related to some measure of need for food aid).

3. Provide greater transparency in commitment performance and ensure the evaluation of the quality of food aid activities.
4. Increase participation of other stakeholders, especially recipient governments.
5. More closely integrate the convention with other international food security arrangements (e.g., reform of the FAO Committee on Food Security).

In addition, the Right to Food principles related to food aid outlined in the Food and Agriculture Organization of the United Nations (FAO) Voluntary Guidelines on the Right to Food (FAO 2005) were discussed and largely supported, particularly as they apply to "do no harm" principles in food aid activities.

Since the Berlin Conference, discussion has continued on the possibility of an International Food Assistance Convention, particularly supported by the European Union, with interest shown by Canada. Particular challenges relate to the quantification of commitments when applied to food, vouchers, and cash, particularly if the convention is to retain its commitment to a quantity of food rather than a quantity of money.

A Further Development?—The Integration of a Human Rights Approach

Some of the member states and the TAFAD group have shown their support for a stronger human rights orientation for a new convention. The Human Right to Adequate Food, already recognized in the United Nations International Covenant on Economic, Social, and Cultural Rights, has received increasing attention in recent years. This Right to Food orientation would apply at two levels—ensuring that food aid activities do not interfere with the right to food of either the recipients or adjacent populations, and ensuring that the overall commitment is sufficient to support meeting the need for food aid. A human rights approach to food aid would include the following aspects:

Respecting and protecting the right to food

The principles of respecting and protecting the right to food do not carry any particular resource transfer elements. Practically, they involve such issues as ensuring that food aid does not negatively impact local food markets, that rations provided are nutritionally and culturally adequate, and that no groups are discriminated against in food aid distribution. Integrating this approach into the convention would require paying much more attention to the way in which food aid activities are implemented. This echoes the other calls for more attention to assessing the quality of implementation of food aid activities. It requires that the convention develop much greater capacity to assess the quality of food aid activities in its mandate, possibly suggesting greater involvement by the World Food Programme or the FAO.

Supporting the fulfillment of the right to food

While the right to food places the primary obligations on the state in which food insecurity is taking place, it also stipulates that, if the recipient state can demonstrate that it is unable to fulfil the right to food in emergency situations, it should appeal to the international community. The international community then has a responsibility to support national governments to ensure that the right to food is realized for those facing hunger. Monitoring and meeting this responsibility could be one of the functions of the convention and would serve as a response to the call to make the total convention commitment level meaningful.

Conclusion

The renegotiation of the Food Aid Convention will provide an important opportunity to strengthen the global food security architecture as we enter an era of increasing instability in the world's food systems. It provides a tangible means of contributing to such internationally recognized goals as the first Millennium Development Goal. But the Food Aid Convention is currently like a T-shirt in a blizzard—better than nothing but certainly not up to the task. Nevertheless, the basic structure of the convention is sound, and, with appropriate and modest changes, it could become an important symbol of the international community's ability to work together to address common hunger challenges. Whether or not its members rise to this challenge will depend very much on whether political support can be successfully generated, especially in difficult economic times.

Note

1   Beyond the semantic objection of referring to all food aid as wheat, the feat of finding a system to add up so many "apples and oranges" is not insignificant.

Works Cited

Benson, Charlotte (2000). "The Food Aid Convention: An Effective Safety Net." In *Food Aid and Human Security*, ed. E.J. Clay and O. Stokke. London: Frank Cass.

Cohen, Marc J., and L. Weingärten (2007). "Food Aid: Exploring the Challenges." Conference report from Food Aid: Exploring the Challenges. Berlin: Federal Ministry for Economic Cooperation and Development. 2–4 May.

Food and Agriculture Organization of the United Nations (2005). "Voluntary Guidelines to Support the Progressive Realization of the Right to Adequate Food in the Context of National Food Security." Rome.

Hoddinott, John, Marc Cohen, and Christopher Barrett (2008). "Renegotiating the Food Aid Convention: Background, Context and Issues." *Global Governance* 14, no. 3: 283–304.

Sphere Project (2004). "Humanitarian Charter and Minimum Standards in Disaster Response." Oxford: Oxfam Publishing.

United Nations (1986). "Food Aid Convention, 1986." London. 13 March.

——— (1995). "Food Aid Convention, 1995." Geneva. 1 April.

——— (1999). "Food Aid Convention, 1999." London. 13 April.

CHAPTER 8

From Food Handouts to Integrated Food Policies

Frederic Mousseau

The increase in food prices in 2007 and 2008 exposed a pre-existing global food crisis that was already affecting about 850 million undernourished people. Food-price increases have exacerbated the dire living conditions faced by hundreds of millions in the poorest countries. Even before the food riots in early 2008, some 16,000 children died every day from hunger-related causes—one every five seconds (Black, Morris, and Bryce 2003).

The situation is getting worse now that the number of undernourished people has increased to over 1 billion. Food prices have become more volatile due to their increasing dependency on uncertain oil markets, while climate change increases the frequency and intensity of disasters, resulting in shocks to both food production and supply.

Policy shifts are required to reliably tackle world hunger through sound agriculture and trade policies. Meanwhile, governments and aid agencies must take concrete action to effectively meet the food needs of the poor in developing countries and protect them against disasters and market volatility. But what should these concrete measures look like? And who should do what?

How Is Hunger Addressed Today?

International emergency interventions are required in order to save lives and protect and restore livelihoods in crisis situations, such as Darfur, where governments lack the capacity or political will to do so. However, food relief does not only take place in contexts of war or major disasters. The World Food Programme (WFP) delivers food aid in some eighty countries, most of them at peace and with stable governments, and only fifteen are conflict zones such as Congo or Afghanistan (WFP 2009).

The WFP and NGOs such as Oxfam are increasingly called upon to deploy relief interventions to help people in extreme poverty who are facing the vagaries of weather or markets. Since the turn of the millennium, all the major food emergencies that triggered international interventions in Africa—Southern Africa in 2002 and 2005, Sahel in 2005, Horn of Africa in 2000, 2002, 2006 and 2008—took place in contexts of chronic hunger and poverty, not war or major disaster. Where millions of people live precariously on the edge of survival, with no access to safety net programs or insurance, and with few savings or assets to fall back on, relatively small economic or climatic shocks can create acute crises. The relief provided is often critical to save lives and protect livelihoods, and these organizations should be enabled to do more. But, despite their commitment and worldwide presence, aid organizations are not in a position to meet all needs. The international aid system is not fit for this purpose; it cannot work at the scale required, and should not, in fact, seek to take on what should properly be the responsibility of national governments in the affected countries.

One of the most dramatic stories concerning the rise in food prices in 2008 was the challenge to the World Food Programme's budget. The WFP was confronted with a 35 percent increase in operational costs due to the higher costs of food and transportation, and had to find an additional US$755 million to maintain its assistance to some 70 million people (WFP 2008). However, these beneficiaries only constitute 8 percent of the total number of undernourished people worldwide.

The WFP eventually got their US$755 million and increased their budget by nearly 50 percent in 2008 to US$5 billion in order to scale up their programs in response to the high food prices. Although this allowed the program and its ten thousand permanent staff to reach 100 million people worldwide, it is still very far from the target figures necessary to tackle world hunger. If the challenge of meeting the needs of the remaining 900 million is to be achieved, a major increase of United Nations and NGO assistance is essential. But would this be enough?

A Global Food Reserve?

We heard a lot in 2008 about an International Food Policy Research Institute (IFPRI) proposal, endorsed by the World Bank, to create a global food reserve consisting of virtual and physical elements (see von Braun and Torero 2008). The "virtual" system would be supported by funding pledged by big exporting countries and would be used to intervene in grain futures markets to discourage speculation when prices are rising. The notion of a virtual reserve is based on the idea that speculation on futures markets for grain and other commodities is a key factor of the price increase, which would be curbed by such a mechanism. The proposed system would be combined with physical stocks to be managed by the WFP at regional level to meet humanitarian needs.

Let us review this proposal, discussing first the matter of physical stocks. Do we need them? The answer is no. If we look at the major food crises of the past decade, including the examples of Malawi and Niger below, it appears that the problem is not a matter of having stocks readily available but rather the availability of financial resources and swift decision making.

In Malawi, following a bad harvest in April 2001, the government requested in July international assistance to help provide the 600,000 metric tons of food needed in the country. Donor countries were apparently skeptical about the severity of the situation and cautious after reported mismanagement of the national food reserves. As a result, they did not meet this request. In April 2002, however, one year after the bad harvest, reports of starvation in some parts of the country galvanized the WFP to launch an emergency appeal, and a massive relief operation finally began with strong support from the main donor countries. It was unfortunately too late for those who died during the lean period in the first months of 2002, when food stocks were depleted and food prices were at their highest level. Instead, Malawi was flooded with food one year after the failed harvest, with serious adverse effects on the country's budget, economy, and agriculture. Mozambican farmers were also seriously affected by the depression of the regional market (Mousseau 2004).

Similarly, the government of Niger and the WFP called for aid in November 2004, following a bad harvest one month earlier. The response from the international community was initially very limited. Four months after its first appeal, the WFP had received only 10 percent of the required funding. The international community largely ignored the situation in Niger until July 2005, at which point expensive emergency nutritional products were routed by air from Europe to treat widespread malnutrition. Regional purchases of food months before could have easily prevented the severity of the situation. Indeed, it would have been possible to organize an appropriate response as soon as food shortages were predicted, providing funds to the government and UN agencies for food imports and distribution, and allowing smooth and organized support from the international community (Mousseau and Mittal 2006).

These two examples demonstrate how the lack of timely decision making transforms food deficits into food crises. They also recall that when there is a food deficit somewhere, because of drought or floods, for instance, there is generally still food available for purchase in the country or the region. After the Asian tsunami in December 2004, only coastal lands were affected, and thus food could be immediately supplied from inland farms and stocks in the case of Sri Lanka, or from other regions in the case of India and Indonesia.

In any case, what is essential in case of disaster is to ensure swift decision making and the availability of funding to address the problem, as opposed to maintaining stocks of prepositioned food. It is now widely acknowledged that it is critical to prioritize food purchases as close as possible from deficit areas in order to support local production and trade. For chronic problem situations in which food assistance is required every year, such as in the case of Ethiopia, it is necessary to ensure predictable, multi-year commitments from donors so that governments or the WFP do not have to launch emergency appeals on an annual basis. This is what is currently happening in Ethiopia, following the establishment of a Productive Safety Net program in 2005.

The other element of the IFPRI proposal concerns the issue of virtual reserves. Do we need them? I suggest three elements of response:

1. Is it realistic? What is the incentive for the main food exporters to intervene in futures markets to keep prices low and reduce the profits of their farmers and traders? How can we realistically expect these countries to make such commitments when it does not seem to be in their interest?
2. We already have a global mechanism, the Food Aid Convention (FAC), that commits donors to provide a minimum of 5 million tons of food aid every year. While the FAC needs to be reformed in order to make it a more effective anti-hunger instrument, it is in place and can be used.
3. The most critical problem is that neither a global mechanism nor the associated proposal of WFP-managed regional physical stocks would contribute to the capacity and resilience of developing countries to face supply shocks. Both mechanisms would maintain and possibly increase developing countries' dependence on the good will of exporting countries and international institutions for their food supply.

Restoring Food Policies

Until 2008, many countries had given up the very idea of running food policies. It was widely believed that global food markets had become larger and less volatile, and that countries were better off liberalizing their economy and buying food on international markets, when necessary, than they were supporting local food production and holding domestic stocks.[1]

This approach was strongly promoted by major donors, many of which are also major cereal exporters and enjoy a dominant position on the global food markets.

Yet, it remains necessary that states bear the responsibility for realizing people's rights to adequate food and livelihoods. Unfortunately, today, hunger is too often ignored, accepted as "given," and left unaddressed with the hope that long-term development will eventually solve the problem. Hundreds of millions thus live daily in hunger, which forces them into poverty and a permanent struggle for survival. Millions do not make it and die from hunger-related causes every year.

Apart from specific crisis conditions, national actors, primarily developing-country governments, must play a far larger role in the solution, with the participation of local civil-society organizations. States should not rely on international relief and food aid to meet the basic needs of their people but rather should take all relevant measures to meet people's ongoing needs and protect them against shocks.

Many governments have been successful in reducing hunger and vulnerability to disasters through sound and comprehensive public policies. For instance, over the last fifteen years, Brazil's comprehensive approach to food security has greatly reduced the prevalence of hunger in the country, with malnutrition in children under the age of five falling from 13 percent to 7 percent between 1996 and 2006 (Brazilian Ministry of Health 2008).

However, not all countries have the capacity or the political will to address the problem on their own. Due in large part to structural adjustment programs and strong pressure from donors, many poor countries have greatly reduced public interventions in social and economic sectors over the past three decades. This has resulted in even greater market volatility and has undermined the capacity of states to alleviate hunger and respond to food crises.

It is now time to rethink food policies in order to deal with hunger at scale, and to seek innovative ways to deal with the problem of market volatility, giving states a key role in ensuring that food is available and affordable to all. States must use a combination of policies and interventions adapted to every context, including, for instance, employment schemes, direct transfers of food or cash to vulnerable groups, food and input subsidies, insurance, credit, food reserves to stabilize prices, trade and fiscal measures to help poor consumers, market regulation, and support for farmers. They must also fix minimum wages and other labour rights at the required level necessary to meet basic needs.

Steps in a New Direction

Reassess the cost effectiveness of food reserves
One of the central arguments made against country-managed grain reserves points to the high cost of these instruments (Rashid, Cummings, and Gulati

2005). Because developing countries' food import costs have increased substantially (13 percent in 2006 and 33 percent in 2007), this argument needs to be reconsidered. Moreover, a reassessment of the cost effectiveness of these instruments must be comprehensive enough to consider the direct costs of managing and holding stocks versus the cost of not having such instruments in place (e.g., hunger and malnutrition toll, costs of safety nets, nutrition and food aid programs).

Link public procurement and storage of food to agricultural policies

Public procurement, along with the implementation of ceiling and floor prices, can be used to protect consumers and producers against market volatility. This is how Europe and Northern America have developed their agriculture over the past fifty years. IFPRI shows that the stability of prices ensured by parastatals in Asia has mitigated risks and given farmers some degree of certainty in allocating their land in favour of the crops for which prices are guaranteed. This has resulted in a positive impact on agricultural development and substantially increased economic growth in several countries (see Rashid et al. 2005, 35–39).

Active participation

The active participation of farmer organizations and civil society organizations must be favoured in the governance and management of parastatals through concrete arrangements and institutional reforms that aim to address the concerns of cost-effectiveness, corruption, and management of public mechanisms. Oxfam research found this to be a far better option for Malawi's Agricultural Development and Marketing Corporation (ADMARC) than the privatization advocated by some donors (Nthara 2002). Concrete arrangements for the participation of farmer and civil society organizations will also improve the accountability of food policies and institutions toward the poor and the hungry.

Innovation in community storage

It is important to reinvest in community storage, such as grain banks, and innovative ways of storing food, such as warehouse receipts or warrantage involving both the public sector and the private sector through financial institutions. Cereal banks and warrantage allow decentralized, community-based systems of food management, intended to protect farmers and consumers against market fluctuations. Cereal banks buy grain from farmers at the harvest time when prices are low. The food is stored until the lean season comes, along with higher prices. The food is then sold below market prices but with a margin to cover management costs and future purchases. The warrantage system, in place in Sahel since the 1990s, provides a similar function but is operated by farmer groups and offers credit to farmers with the support of financial institutions. The farmers sell the food at harvest time, which is kept in storage and sold a few months later when prices are higher. Farmers then obtain the additional revenue generated. This acts as a market regulating tool and serves as an important safety net, especially for households facing other vulnerabilities. The effectiveness of these instruments has been uneven in the past, often limited by lack of cash flow and the vulnerability of these institutions to market fluctuations. The lack of durable financial and management support and poor institutional backup have also undermined the sustainability of these instruments.

Think regionally

National reserves may not be appropriate anymore in economically integrated regions. The Southern Africa Development Community announced in 2008 the constitution of a 500,000 metric ton food reserve, to be procured in the region. Countries in other regions, such as West Africa, could consider adopting such a model, which may be more appropriate than national reserves in the context of open regional markets. However, the development of regional reserves needs to be preceded by the development of regional agricultural and food policies, because food reserves cannot be managed independently and must be handled with flexibility according to prices and production levels.

Innovative food procurement

Developing countries must also consider innovative ways to procure food for public interventions such as what is happening with the Southern Africa Futures Exchange, where a country like Malawi is able to negotiate and secure the procurement of food at pre-agreed prices on the regional food exchange.

Conclusion

Though the food crisis is a global one, it is essential to recognize that it must be primarily addressed at the local, national, and regional levels rather than through the creation of new global mechanisms.

To be successful, investment in agriculture requires policies designed and implemented by states, with strong participation of all parties, including civil society organizations, unions, farmer groups, fishing communities, and so on. The 2008 food crisis made it clear that the international aid and cooperation system must revisit old paradigms and support the restoration of true food policies. The international community should actively support governments to assume their responsibilities and put in place adequate public mechanisms and interventions. In order to do so, donors and aid agencies must stop pressing for rapid liberalization of economies, shift away from an emphasis on the delivery of food and projects, and provide more support to local mechanisms and institutions, including through direct budget support to governments when appropriate.

Note

1   As an Overseas Development Institute paper put it in 2003, "greater integration into the international market would reduce the variability of food prices" (Anderson and Slater 2003).

Works Cited

Anderson, Edward, and Rachel Slater (2003). "Food Security in Indonesia." In *Food Security and the Millennium Development Goal on Hunger in Asia* (ODI Working Paper, No. 231), ed. G.J. Gill, J. Farrington, E. Anderson, C. Luttrell, T. Conway, N.C. Saxena, and R. Slater. London: Overseas Development Institute.

Black, Robert, Saul Morris, and Jennifer Bryce (2003). "Where and Why are 10 Million Dying Every Year?" *The Lancet* 361, no. 9376: 2226–34.

Brazilian Ministry of Health (2008). "Pesquisa Nacional de Demografia e Saúde da Criança e da Mulher." Brasilia. http://bvsms.saude.gov.br/bvs/pnds/index.php.

Food and Agriculture Organization of the United Nations (2008). "Number of Hungry People Rises to 963 Million." Rome. 9 December.

Mousseau, Frederic (2004). "Roles and Alternatives to Food Aid in Southern Africa: A Review of the Southern Africa Food Crisis." Oxford: Oxfam Great Britain. http://www.sarpn.org.za/documents/d0000998/P1121-Roles_and_alternatives_to_food_aid_Mousseau_2004.pdf.

Mousseau, Frederic, and Anuradha Mittal (2006). "Sahel: A Prisoner of Starvation? A Case Study of the 2005 Food Crisis in Niger." Oakland, CA: Oakland Institute. http://www.oaklandinstitute.org/pdfs/sahel.pdf.

Nthara, Khwima (2002). "What Needs to be Done to Improve the Impact of ADMARC on the Poor." Blantyre, Malawi: Oxfam Great Britain. http://povlibrary.worldbank.org/files/15033_oxfam_phase2.pdf.

Rashid, Shahidur, Ralph Cummings Jr., and Ashok Gulati (2005). "Grain Marketing Parastatals in Asia: Why Do They Have to Change Now?" Discussion Paper No. 80. Washington, DC: International Food Policy Research Institute. http://www.ifpri.org/divs/mtid/dp/papers/mtidp80.pdf.

von Braun, Joachim, and Maximo Torero (2008). "Physical and Virtual Global Food Reserves to Protect the Poor and Prevent Market Failure." IFPRI Policy Brief, no. 4. Washington, DC: International Food Policy Research Institute.

World Food Programme (2008). "UN Press Statement: A Unified United Nations Response to the Global Food Price Challenge." Rome. 29 April.

———— (2009). "About—Fighting Hunger Worldwide." Rome. http://www.wfp.org/about.

CHAPTER 9

The Uses of Crisis

Progress on Implementing US Local/ Regional Procurement of Food Aid

Gawain Kripke

This chapter reviews recent events and progress in the advocacy effort to reform US food aid programs to enable local and regional purchase of food aid. It provides details on several venues in which food aid reform was considered and debated and notes that progress has been limited. However, the 2008 food price crisis has changed the external context for food aid and helped unblock policy debates over local and regional purchase in the US Congress. Food aid is a sliver of international assistance, making up approximately 3–5 percent of overall official development assistance (ODA) in recent years. While these US$3–4 billion in aid are a critical lifeline to people in desperate need, food aid has been an area of disinterest and neglect in the broader field of development.

Although recognized as a necessary measure, food aid is often ignored, considered a crude, and ideally temporary, intervention. Delivering food to beneficiaries is a logistical and funding challenge, but is not very interesting as a development strategy. It is viewed more as a short-term palliative than a long-term solution.

In this way, food assistance seems out of date in modern development discourse, which often concerns itself with issues of empowerment, entitlements, livelihoods, and power. At worst, food aid could actually increase dependency and depress the livelihoods of poor people themselves.

On the other hand, the importance of food assistance as a humanitarian intervention is disproportionate to its share of ODA. Food aid literally saves lives, which has an enormous welfare value. Since the large majority of food aid is provided on an "emergency" basis to people in acute need, this aid is among the most tangible, direct, and measurable transfers from donors to poor and vulnerable people. The impact on beneficiaries is very high, even if not long lasting. For young children, however, adequate nutrition can prevent irreversible harm: children who experience malnutrition between birth and two years old can suffer a range of long-term impacts including physical stunting, increased risk of disease, reduced cognitive function, and lower economic productivity (Ruel and Hoddinott 2008).

In recent years, food aid programs have become the subject of controversy and critique both on the international stage and within the United States. This debate has been sparked in large part by a movement emerging among development stakeholders to reform and improve the execution of food aid programs.

Critique of Food Aid Programs

Existing food aid programs have been the subject of a range of critiques in recent years. The primary target of many of these critiques is the United States. As the donor of close to half of all international food aid, the United States plays a major role in supporting and maintaining existing food aid programs. The United States is the largest contributor to the UN World Food Programme (WFP), the multilateral food aid agency. In addition, the United States uses food aid to support a variety of bilateral programs through various NGOs, contractors, and governments. However, the United States makes virtually all of its food aid contributions in the form of food commodities procured on US markets and largely shipped on US-flagged vessels.

The critiques of US policy have overlapping, but distinct themes: its potential to disrupt markets, problems associated with its donor-driven nature, and inefficiencies.

Among the most damaging critiques of food aid is that it can displace commercial food markets at both the international and domestic levels. There is little argument that some forms of food aid displace commercial imports. Indeed, this was one of the purposes of some first-generation food aid programs: to support governments by reducing their need to purchase and import food (US General Accounting Office 1995). Over time, improved targeting has helped to reduce the commercial displacement caused by food aid. Nonetheless, critics argue that food aid continues to displace commercial food imports and that, even worse, it can depress local food production and marketing. This is a very problematic critique, since one of the primary purposes of food aid is to improve food security. But if food aid floods local markets and depresses prices, local farmers could lose income and, consequently, their incentive to continue producing food. For the longer term, this undermines local food production and longer-term food security (Clay, Riley, and Urey 2004; Barrett and Maxwell 2005; Lavy 1990; Abdulai, Barrett, and Hoddinott 2005).[1] This argument has a clear logic, although it is supported more commonly with anecdotal evidence than with rigorous analysis (Thurow and Kilman 2003; Saunders 2005; and Oxfam GB 2001).[2]

Some analyses show that poorly timed or over-sized food aid interventions can contribute to depressed prices. For example, Oxfam made the case that in 2002 and 2003, food aid donors overreacted to a projected 600,000 metric ton food deficit in Malawi and sent close to that amount in food aid. However, commercial and informal importers also brought in an additional 350,000–500,000 metric tons, leaving Malawi flooded with large carry-over stocks. Maize prices subsequently dropped from US$250 per metric ton to US$100 per metric ton in the course of a year. Local production of maize, cassava, and rice fell markedly, and estimated losses to the Malawian economy were approximately US$15 million (Mousseau 2004).

Broader economic studies are inconclusive about the extent to which food aid causes a disincentive to local production. Most studies of food aid impacts are conducted at a national or global level, using aggregated data (Clay et al. 2004). This hides impacts in local markets, where price depression and displacement are more likely, especially in the fragmented markets typical of many countries receiving food aid.

There is some evidence that the use of food aid correlates with long-term dependence on food imports. As food aid declines, commercial imports tend to take its place, rather than local or national food production (FAO 2004, 17).

Another critique of food aid is that it is donor-driven, rather than driven by the needs and interests of beneficiaries. The flow of food aid tends to rise when donors have food surpluses and when food prices decline, and tends to decline when food prices rise. This creates a pro-cyclical pattern, which means less food aid is available when it is needed more. Food aid volumes are closely correlated to carry-over surpluses; when surpluses rise, so does food aid. This pattern was especially notable in 2008 as food prices climbed to record levels, driving large numbers of people into food insecurity and causing dramatic civil disturbances and even political instability (including changed governments in food import-dependent Mauritius and Haiti). At the same time, the WFP faced a shocking budget shortfall in food needed to meet commitments, at one point claiming a US$755m budget deficit.

Finally, food aid given in commodity form is inefficient. Critics point out that food aid donated as commodities can take much longer to deliver than food aid purchased closer to the targeted beneficiaries. This is an important consideration when weeks or even days can mean the difference between life and death during humanitarian emergencies. In the US system, it can take months from the date of a procurement order until food aid is actually delivered to port. US emergency shipments experienced a median lag of nearly five months in 1999–2000, due to bureaucracy and cumbersome procurement restrictions—and, of course, the need to ship food over long distances (Barrett and Maxwell 2005).

Sending food over long distances, with restrictive procurement and shipping requirements, means that funds are spent on bureaucracy, processing, and shipping rather than on the food and its distribution. In 2004, the Organization for Economic Co-operation and Development produced a study based on a review of thousands of food aid transactions. The study found that shipping food from donor countries is 33 percent more expensive than buying it from a third-party country (usually closer to the destination) and 46 percent more expensive than buying it locally in the destination country (Clay et al. 2004).

US food aid is notably inefficient in this regard. US commodity suppliers enjoy an 11 percent premium above commercial prices for food aid purchases. In addition, because US law requires that 75 percent of food aid shipments use US flag-carriers, a 78 percent premium is paid on food aid shipments. By donating food that is US-sourced and US-shipped, US taxpayers lose more than half the value of their food aid dollar in costs by the time food aid reaches the destination port (Barrett and Maxwell 2005).

Making food aid contributions available as cash rather than food commodities is a policy reform that would fully or partially address each of these critiques. This simple step would help address the commercial displacement argument—particularly if the food were procured in local markets, thus contributing to local market demand and helping to incentivize continued food production. Shifting to cash rather than commodities would put donor suppliers and shippers on equal footing to compete for the food aid contracts, likely lowering costs by removing monopolistic rents created by "buy America" policy mandates. In the process, the critique that food aid serves the donor's interest would be mooted.

Reforming US food aid policy so that a portion, or all, could be contributed as cash rather than as food commodities has emerged as a goal for many stakeholders in the field of food security and food aid in the United States. However, defenders of existing arrangements in the US food aid programs make a political argument: tying food aid to the purchase of commodities creates a convergence of interests between humanitarian goals and the commercial interests of donor country agriculture producers, processors, distributors, and shippers. This coalition of interests is powerful and underlies the resilient political support for food aid programs. Without this coalition, defenders argue, support for food aid would wane and funding decline.

A Record of Frustration

Although critiques of food aid emerged in recent years, US policies remained largely unchanged. A combination of political inertia and interested lobbying made reform of US food aid programs impossible for several years; they remained protected by the wall of legislative inertia embodied in the "Farm Bill," the primary legal authority for the US food aid program. The Farm Bill is an omnibus legislative vehicle that includes agricultural subsidies, rural development programs, and domestic and international anti-hunger initiatives, and authorizes these programs for discreet amount of time—usually five years. During that time, the Farm Bill can be amended, but it is a significant political hurdle to re-open the bill. A version of the Farm Bill was enacted in 2002, after which there was little political appetite in Congress to revisit the issue of food aid—at least until the bill expired in 2007.

In 2005, President George W. Bush emerged as a leading advocate for reform of the US food aid program. As part of his annual budget proposal to Congress, President Bush recommended a reform to provide up to 25 percent of food aid budgets as cash contributions rather than as in-kind commodity contributions. This proposal came as part of the president's annual budget proposal to the US Congress. However, with little public debate, Congress rejected this proposal. The House Agriculture Appropriations Committee was particularly forceful, noting, "we are pleased that the Committee rejected an ill-advised administration proposal to move $300 million of international food aid to USAID, rather than continuing to fund it through the PL 480 program. The USAID funds would be used to buy food abroad, instead of American commodities, thus undermining the historically broad support for international food aid in this country" (US House of Representatives, Agriculture Appropriations Committee 2005).

President Bush repeated the proposed reform in his budget requests of 2006 and 2007, however, each time, Congress rejected it (Hanrahan 2007).

While reform of food aid sputtered in Washington, another forum for debate and negotiation gained traction in Geneva. Trade negotiations as part of the World Trade Organization (WTO) Doha Round offered a chance for substantive change to food aid programs. Launched in late 2001, the Doha Round was intended to introduce new, tougher rules on trade distortions in agriculture. Under this rubric, food aid was defined as an "export subsidy," subject to disciplines under trade rules. In previous trade negotiations (the Uruguay Round), food aid was mentioned under the heading of export subsidies in the text, with some advisory language. But the WTO did not (and still does not) have enforceable rules on food aid. Reforming agriculture generally, and export subsidies in particular, was considered important "unfinished business" from the Uruguay Round trade negotiations that the Doha Round would tackle.

After fitful progress, in 2004, new rules for food aid were explicitly nominated for negotiation:

> Provision of food aid that is not in conformity with operationally effective disciplines to be agreed. The objective of such disciplines will be to prevent commercial displacement. The role of international organizations as regards the provision of food aid by Members, including related humanitarian and developmental issues, will be addressed in the negotiations. The question of providing food aid exclusively in fully grant form will also be addressed in the negotiations. (WTO 2004)

In the context of trade negotiations, the arguments around food aid had little to do with humanitarian or efficiency critiques, but everything to do with the "commercial displacement" that it might cause. For example, export competitors such as New Zealand complained that the United States was shipping milk products to Central America as food aid (WTO 2003). South African maize millers publicly complained that US food aid in Southern Africa was displacing commercial trade (Louw 2004).

The United States resisted new rules, arguing that food aid, as a development and humanitarian instrument, was beyond the competencies of the WTO. However, while the United States delayed offering a substantive proposal, negotiating partners persisted and, over time, submitted a series of progressively more detailed proposals for rules to discipline food aid. The United States agreed to language at the Hong Kong WTO Ministerial Conference that concretized some of the negotiating concepts, including:

- a "safe box" for bona-fide food aid in emergency situations that would not be subject to challenge or discipline under the WTO;
- a commitment to "ensure elimination of commercial displacement;" and
- rules on in-kind food aid, monetization, and re-export of food aid (WTO 2005).

In 2006, the United States offered a substantive proposal at the WTO that adopted these concepts but preserved a large amount of policy space for existing food aid programs and practices, including delivery of in-kind food aid and monetization of food aid (USTR 2006). Proposals from other countries came as well, including Canada, the Cairns Group of agriculture exporters, the EU, the African Group jointly with the least developed countries group, and others. The majority of these sought to add increased specificity and rigour to the rules. These proposals pushed for all food aid to be provided in grant form only, rather than as concessional sales or loans. They called for significant restrictions on monetization of food aid to eliminate or narrow its use. They attempted to clarify and specify how and who could declare an emergency, thus qualifying food aid under the "safe box."

The Doha Round negotiations on food aid generated some political energy around reform—both in favour and opposed. Lobbyists representing US NGOs and private-sector beneficiaries of the food aid program made visits to Geneva to meet with WTO delegations and argue against the reforms. They sent letters and papers articulating opposition to WTO disciplines. They also worked with some WTO delegations—notably Mongolia—to propose less rigorous rules (Coalition for Food Aid n.d.; *Inside U.S. Trade* 2003).

Although some progress was being made in the negotiations around food aid, and some convergence could be observed in negotiating positions, the broader Doha negotiations were not going well. Following the Hong Kong WTO ministerial conference in late 2005, little substantive progress was made in subsequent negotiating sessions, such that, by 2009, the Doha Round negotiations were widely viewed as deadlocked and unlikely to advance in the near future.

However, just as the Doha Round negotiations began showing signs of stalling, negotiations around a new Farm Bill began to take shape.

As mentioned above, the 2002 Farm Bill was scheduled to expire in 2007, requiring Congressional action to extend the many programs under this law. In general, food aid plays a very small role in Farm Bill debates, making up only a small portion of the overall spending under the legislation: approximately US$1.5–2 billion of an annual total of about US$60 billion. The bulk of the Farm Bill is composed of domestic anti-hunger programs and commodity subsidies, with significant environmental conservation and rural development programs included. Food aid is never a headline issue in the Farm Bill and is often not publicly debated at all.

To launch the Farm Bill debate, then US Secretary of Agriculture Mike Johanns proposed draft legislation on 31 January 2007. The Johanns proposal would have made up to 25 percent of US food aid budgets available as cash contributions rather than requiring US procurement and shipping. As the Bush administration argued, "local and regional purchases will be used judiciously where the speed of the arrival of food aid is essential. The Administration will be better equipped to deal with emergencies if our tools include cash that can be used to provide immediate relief until US commodities arrive or to fill in when there are pipeline breaks," and further, "the principal reason for the proposal is to save lives. USAID's conservative estimate is the authority [to procure food aid locally or regionally] could feed at least one million additional people for

6 months and could save at least 50,000 lives in acute emergencies" (USDA 2007, 81–82).

Although the Bush administration made a clear call for reform in the food aid program, there was little response in the Congress. In particular, there was no apparent support for food aid reform among the members of the House and Senate Agriculture Committees, which were charged with rewriting the Farm Bill.

In May 2007, the House Agriculture Committee chairman introduced a draft Farm Bill that effectively extended the status quo in food aid programs through 2012. Debate on the Farm Bill in the Agriculture Committee and on the floor of the House of Representatives did not address food aid. Only one serious effort was made to reform the food aid program during the House consideration of the Farm Bill: Representative Earl Blumenauer (D-OR) requested to offer an amendment to the Farm Bill that would have made US$100 million available annually for a pilot project for local and regional purchase of food aid. However, Rep. Blumenauer was denied the opportunity to offer the amendment during consideration of the Farm Bill by the powerful Rules Committee, which is controlled by the Congressional leaders.

During Senate consideration of the Farm Bill, the food aid program emerged for a slightly more substantive debate. The draft legislation offered by the chairman of the Senate Agriculture Committee included a "pilot project" for local and regional purchase of food aid. This pilot was authorized for US$60 million over four years, an average of US$15 million per year. This is a tiny fraction of the total US food aid budget of approximately US$2 billion annually. Even still, the pilot was heavily conditioned and stipulated in the legislation—with requirements such as:

- at least one project carried out jointly with a project funded through grassroots efforts by agricultural producers through eligible United States organizations;
- projects in both food surplus and food deficit regions, using regional procurement for food deficit regions; and
- projects in diverse geographical regions, with most, but not all, projects located in Africa.

This level of micro-managing and conditions was extraordinary and reflected a variety of special interests in the food aid program—each seeking special attention to their concerns. More generally, given the small amount of funding but heavy restrictions, the program appeared unlikely to produce robust and positive results.

When the Senate Agriculture Committee met to debate the Farm Bill, Senator Pat Roberts (R-KS) raised an objection to the food aid pilot, arguing that funding for a pilot should not derive from the existing food aid program. He proposed to leave the pilot program intact, but remove funding for the project, which could instead be funded out of other accounts. This measure could easily have meant that the pilot would never be implemented. No vote was taken and no final decision made during the Senate Agriculture Committee consideration.

The pilot project disappeared from the Senate Farm Bill when the legislation was introduced for debate on the Senate floor, but, it was restored by the time the Senate finalized the bill in December 2007 (US House of Representatives 2007).

A long and difficult "conference" followed, during which House and Senate negotiators hammered out a compromise version of the Farm Bill. A final compromise Farm Bill did not emerge until May 2008. This version did include the pilot project and became law.

Although the pilot project was the first explicit support that Congress had provided for local and regional purchase of food aid, it was disappointing to reform advocates who had hoped for a more substantial reform.

A Food Price Crisis Emerges

Food commodity prices on international markets started a steady upward climb beginning in 2002. Then, in 2007, prices began accelerating through the summer of 2008.

Because high prices are not just a localized problem but affect everyone, the impacts were widespread, affecting rural and urban, poor and rich alike. Acutely vulnerable people could be found in both rural and urban settings. In dozens of countries, high food prices generated civil unrest, including protests and strikes. In at least two countries (Haiti and Mauritania) the food price inflation contributed to the downfall of the government. These protests and disturbances captured media attention, and major news outlets dedicated significant resources to covering the matter, even setting up special websites to collect stories and information and offer thematic coverage (BBC 2008; *Financial Times* 2008).

US affluence means that the average US household spends about 10 percent of its income on food, so even dramatic price increases have only a modest impact on US populations. However, the media coverage of the price inflations—and particularly the political unrest and "food riots" in some countries—brought the issue to the attention of policy-makers in Washington. In April 2008, President Bush drew down approximately US$200 million in emergency food reserves to support ongoing international food aid programs. In May, he requested an emergency package of US$770 million to respond to the crisis. This amount included funds for a variety of related purposes, including funds to preserve price parity to existing food aid programs, essentially a budget "top-up" to maintain food aid volumes even as the price of food increased (Kunder 2008).

Reform Becomes Real

In response to President Bush's emergency package for the food crisis, Congress raised the bid and enacted larger packages of assistance to address the food price crisis. Congress enacted approximately US$1.8 billion for food crisis-related international assistance, including substantially more funding for food aid and for development and disaster assistance to "alleviate world hunger" (US House of Representatives, Committee on Appropriations 2008). The Congressional funding package included US$200 million for a new "Food Security Initiative," which was meant to provide support for agriculture research and development activities. Of this amount, however, Congress specified that US$50 million was to be used for "local and regional purchase" (US House of Representatives 2008).

All told, the emergency funding package included US$125 million in funds from various sources for local and regional purchase programs. The Congressional emergency package was enacted and became law on 30 June 2008 and was non-controversial. This US$125 million for local and regional purchase in the emergency funding package, provided over approximately eighteen months, contrasts starkly with the reform in the Farm Bill, which provides US$60 million over five years.

Neither US$60 million nor US$125 million represents a dramatic shift in US food aid budgets, representing less than 10 percent of the overall food aid budget. However, the trend is clear that after years of inaction, policy-makers in Washington are taking steps to reform the US food aid program to include local and regional purchase.

Considerations and Prospects

What does the US$60 million/five-year Farm Bill pilot project mean? What does the subsequent appropriation of US$125 million over one year for local and regional purchase signify?

While they both represent progress toward a more flexible and effective food aid program, the difference between these measures is worth noting. The Farm Bill pilot project was enacted with great difficulty and despite opposition from various sources.

The funding for local and regional purchase in the emergency food crisis package was enacted very quickly, with little debate and no significant political opposition.

The difference was a crisis; the Farm Bill was crafted before the food crisis emerged as a major media and public concern. Once the crisis attained a high profile, a new political will crystallized to take measures that would respond to it. This included providing funding for new mechanisms to address food security.

There are some legislative peculiarities about the emergency package that facilitated the expansion of local and regional purchase in US food aid. First, the legislation did not run through the House and Senate Agriculture Committees, which are notoriously protectionist about US agriculture. Sidestepping these bodies facilitated a dramatic step away from the status quo, compared to the pilot program the Agriculture Committees included in the Farm Bill.

Second, the local and regional purchase provisions included in the emergency package do not draw funding from the Farm Bill budget and so do not directly threaten the interests supporting the status quo.

That said, the emergency package does represent a break from the near stalemate in discussions around food aid and local and regional purchase. The fact that the legislation passed with no significant debate or opposition to the local and regional purchase is remarkable. And, according to key Congressional staff, the significance of this is real and represents a sea change in the Congressional attitude toward the issue. According to one influential staff member on the powerful Appropriations Committee, "the ability to make progress can be attributed to a crisis, But there's a growing awareness in this Congress... to focus on long-term development" (personal communication, 12 January 2009).

Conclusion

Reform of food aid has been debated for several years, with a growing chorus of academics and advocates arguing for changes to improve efficiency, effectiveness, and expand the benefits of the program by instituting local and regional purchase of food.

Supporters of reform made little progress for years and managed to put in place only a small, symbolic pilot project in the 2008 Farm Bill. The matter might have been concluded until the next Farm Bill in 2012, except for a crisis.

In 2008, food prices accelerated and created a sense of crisis in the media and among policy-makers in Washington. A large package of emergency funding assistance was compiled and included significant new funding for local and regional purchase of food aid.

What can we learn from this? Before claiming any real conclusions, it is important to note that all of these events and actions are still relatively fresh, so it is too early to interpret this recent history. There is still a lot to learn about what has happened and more to observe on how events and initiatives unfold.

That said, it is possible to make some observations that may help inform future critics and advocates.

First is that policy reform is a lumpy process, with fits and starts, and, even when successful, it may not be orderly. For example, a years-long advocacy effort, tracked to a regular legislative process, produced disappointing results, but an ad hoc, emergency response generated significant reform.

Second is that the latter reform is probably informed by the debate and knowledge gained from the former. The advocacy and debate grounded in ideas may not have impacted the targeted policy-makers, but other policy-makers were influenced and carried forward a reform using other means.

A metaphor of a dam with a reservoir can serve to illustrate these lessons. Think of the dam as the policy process. Behind it is a reservoir that is growing full of research, analysis, and idea advocacy; building pressure for reform.

Good dam management would release the growing reservoir in orderly, regular intervals, sending water—new policy—down the river and releasing pressure. But a political process is rarely that organized and can become obstructed by interests. Instead, the pressure for reform will build and may eventually be relieved by bursting through cracks in the dam. Or the water may overflow the dam if the reservoir is full to capacity.

The lesson in this is to recognize that ideas, research, and analysis do contribute to reform, but that the processes by which the reform will occur are likely to be irregular and sometimes require a catastrophic event. Creating the pressure for reform can be a long exercise and very gradual. It can be frustrating to practitioners to find that regular process is unsuccessful. It is useful to keep this in mind and to seek opportunities in crises to advance existing reform agendas.

Notes

1 In some cases food aid can actually help to stimulate local agricultural production. Under some conditions, farmers accessing food aid can reduce their need to spend money on food, which permits investments in productive capacity (such as farm tools) and allows them to access credit. In addition, food aid can improve health and reduce illness, improving labour supply—a critical factor for developing country agriculture. Food aid can also enable agricultural producers to expend limited resources on measures aimed at increasing agricultural production, such as pesticides or fertilizer.

2 It should be said that measuring the depressing impact of food aid on local production would be very difficult, given that food aid often represents a very small portion of overall food markets, and that isolating food aid as a variable could prove complicated.

Works Cited

Abdulai, Awudu, Christopher B. Barrett, and John Hoddinott (2005). "Does Food Aid Really Have Disincentive Effects? New Evidence from Sub-Saharan Africa." *World Development* 33, no. 10: 1689–1704.

Barrett, Christopher B., and Daniel G. Maxwell (2005). *Food Aid after Fifty Years: Recasting Its Role*. New York: Routledge.

British Broadcasting Corporation (2008). "The Cost of Food." *BBC News*. 30 October.

Clay, Edward, Barry Riley, and Ian Urey (2004). "The Development Effectiveness of Food Aid and the Effects of Its Tying Status," DCD/DAC/EEF(2004)9, section 99. Paris: OECD Development Assistance Committee.

Coalition for Food Aid (n.d.). Reports and communications of the Coalition for Food Aid. Washington, DC.

Financial Times (2008). "The Global Food Crisis." *Financial Times*. http://www.ft.com/foodprices.

Food and Agriculture Organization of the United Nations (2004). "The State of Agricultural Commodity Markets 2004." Rome.

Hanrahan, Charles (2007). "International Food Aid and the 2007 Farm Bill." Washington, DC: Congressional Research Service.

Inside U.S. Trade (2003). "'Parallelism' with Subsidies Seen as Beneficial for Export Credits." Washington, DC. 8 August.

Kunder, James R. (2008). "U.S. Response to the Global Food Crisis: Humanitarian Assistance and Development Investments." Statement before the House Committee on Agriculture. Washington, DC. 16 July.

Lavy, Victor (1990). "Does Food Aid Depress Food Production? The Disincentive Dilemma in the African Context." Washington, DC: World Bank.

Louw, Liesl (2004). "Food Fight Flares over US Aid." *News24.com*. 6 February.

Mousseau, Fred (2004). "Roles of and Alternatives to Food Aid in Southern Africa." Oxford: Oxfam Great Britain.

Oxfam Great Britain (2001). "The Impact of Rice Trade Liberalisation on Food Security in Indonesia." Oxford.

Ruel, Marie, and John Hoddinott (2008). "Investing in Early Childhood Nutrition." IFPRI Policy Brief 8. Washington, DC. November.

Saunders, Doug (2005). "Food Aid Exposes the West's Uncharitable Charity." *The Globe and Mail*. 15 January.

Thurow, Roger, and Scott Kilman (2003). "Seeds of Discord: U.S. Food Aid Sparks a Cycle of Dependency for Farmers, Recipients—Sending Crops, Not Cash, Eases American Gluts, Ignores Local Surpluses—A Pitch from Raisin Growers." *The Wall Street Journal*. 11 September.

United States General Accounting Office (1995). "Food Aid: Competing Goals and Requirements Hinder Title I Program Results." GAO/GGD-95-68. Washington, DC.

United States Department of Agriculture (2007). "Farm Bill Proposals." Washington, DC. http://www.usda.gov/documents/07finalfbp.pdf.

United States House of Representatives (2007). "Farm, Nutrition, and Bioenergy Act of 2007." H.R. 2419. Washington, DC.

—— (2008). "Explanatory Statement to Accompany House Amendment #3 – Relating to Consideration of the Senate Amendment to H.R. 2642 – Supplemental Appropriations Act, 2008." Washington, DC.

United States House of Representatives, Agriculture Appropriations Committee (2005). "Agriculture, Rural Development, Food and Drug Administration, and Related Agencies Appropriations Bill, 2006." Washington, DC. 2 June.

United States House of Representatives, Committee on Appropriations (2008). "Emergency Supplemental: Iraq, Afghanistan, Veterans, and Workers." Washington, DC. 14 May.

United States Trade Representative (2006). "United States' Communication on Food Aid." JOB/(06)/78. Geneva: WTO Committee on Agriculture.

World Trade Organization (2003). Notes on WTO Committee on Agriculture. Informal communication from WTO staff. 27 March.

—— (2004). "Doha Work Programme—Decision Adopted by the General Council on 1 August 2004." Geneva. http://www.wto.org/english/tratop_e/dda_e/ddadraft_31jul04_e.pdf.

—— (2005). "Doha Work Programme—Ministerial Declaration." WT/MIN/(05)/DEC. Hong Kong. 13–18 December.

PART 3

Longer-Term Ecological Concerns
and Governance Responses

CHAPTER 10

The Impact of Climate Change on Nutrition

Cristina Tirado, Marc J. Cohen,
Noora-Lisa Aberman, and Brian Thompson

Climate variability and change are expected to exert continuing upward pressure on food prices by drastically reducing production in many developing countries, particularly in sub-Saharan Africa and South Asia, which already form the centre of gravity of hunger and malnutrition (Easterling et al. 2007; FAO 2008b). This will mean increased dependence on imported food and greater vulnerability to volatile world market prices for poor food consumers in those regions.

Furthermore, climate change is expected to increase undernutrition through its effects on illnesses, such as diarrhea and other infectious diseases. In addition, expected increases in the frequency and intensity of droughts and floods are likely to have adverse effects on crops and livestock (Metz et al. 2007). According to the Fourth Assessment Report of the Intergovernmental Panel on Climate Change (IPCC),[1] malnutrition may be one of the most important human consequences of climate change (Confalonieri et al. 2007).

This severe challenge to global nutrition comes amid a background of unacceptably slow progress in reducing malnutrition: between 1990 and 2005, the proportion of underweight preschoolers in the developing world only fell from 30 to 23 percent. At that rate, it will not be possible to meet the Millennium Development Goal (MDG) target of halving the preschool underweight prevalence between 1990 and 2015.

Failure to accelerate progress against malnutrition will have high costs indeed. Inadequate dietary intake and disease are the immediate causes of malnutrition. Inadequate food consumption heightens vulnerability to infectious diseases, which, in turn, can keep the body from absorbing adequate food. These immediate causes stem from insufficient access to safe and wholesome food, poor maternal and child-rearing practices, and inadequate access to clean drinking water, safe sanitation, and health services. Ultimately, these factors are embedded in the larger political, economic, social, and cultural environment. Malnutrition accounts for a high share of the global disease burden. Difficult pregnancies and illnesses due to malnutrition cost developing countries an estimated US$30 billion annually. Lost productivity and income resulting from early deaths, poor school performance, disability, and absenteeism raise the yearly total into the hundreds of billions of dollars.

This chapter explores the implications of climate change for nutrition. We note that agricultural activities contribute to climate change but can also play an important role in adaptation and mitigation strategies. We conclude by examining policy options for addressing the links between climate change and malnutrition. We make the case that a human rights–based approach offers the opportunity to embrace environmental and sustainability concerns more explicitly.

Overview of Climate Change: Evidence for and Potential Effects

According to the IPCC, climate variability and change will lead to more intense and longer droughts, particularly in the tropics and subtropics (Trenberth et al. 2007). In addition, the frequency of heavy precipitation events has increased over most land areas. It is very likely that heat waves and heavy precipitation events will become more frequent and that future tropical cyclones will become more intense (Meehl et al. 2007; Trenberth et al. 2007). It is primarily through these impacts that climate change will have negative effects on nutrition: droughts and water scarcity diminish dietary diversity and reduce overall food availability. The risk of flooding of human settlements may increase, from both sea-level rise and increased heavy precipitation in coastal areas. This is likely to result in an increase in the number of people exposed to diarrheal and other infectious diseases.

Global atmospheric concentrations of greenhouse gases (GHGs), carbon dioxide (CO_2), methane, and nitrous oxide have increased markedly as a result of human activities. Continued GHG emissions at or above current rates would cause further warming and induce many changes in the global climate system during the twenty-first century (Meehl et al. 2007). Water supplies stored in glaciers and snow cover are projected to decline, reducing water availability in regions that are home to more than one-sixth of the world's population (Kundzewicz 2007). Widespread retreat of glaciers and ice caps has contributed to sea-level rise (Lemke 2007). According to the fourth IPCC report, sea level will rise by forty centimetres by the 2080s, with 60 percent of this increase occurring in South Asia and 20 percent in Southeast Asia (Meehl et al. 2007).

Vulnerability

Vulnerability to adverse effects of climate change differs by region, population group, and gender. It should be kept in mind that IPCC assessments provide only weak information at the regional level and none on a national basis.

Vulnerable Regions

The regions likely to be adversely affected by climate change are those already most vulnerable to food insecurity and malnutrition, notably sub-Saharan Africa, which may lose substantial agricultural land (Nicholls et al. 2007). These are also the regions that are most vulnerable to food-price volatility.

In seasonally dry and tropical regions, crop productivity is projected to decrease with even small local temperature increases (1–2°C) (Easterling et al. 2007). In Africa, by 2020, between 75 million and 250 million people are projected to be exposed to increased water scarcity. If coupled with increased demand, this will adversely affect livelihoods and exacerbate water-related problems (Boko et al. 2007; Kundzewicz et al. 2007). In much of Africa, agricultural production and access to food are projected to be severely compromised. This would further adversely affect food security and exacerbate malnutrition on the continent.

Coastal areas, especially heavily populated mega-delta regions in South, East, and Southeast Asia, will be at greatest risk of increased flooding from the sea and, in some mega-deltas, from rivers (Cruz et al. 2007). Sea-level rise will increase salination of groundwater and estuaries, resulting in a decrease in coastal freshwater availability for humans and ecosystems (Kundzewicz et al. 2007).

Vulnerable Populations

The most vulnerable populations will suffer earliest and most from climate change, and this should be addressed in a way that is fair and just, cognizant of the needs and risks faced by the vulnerable groups, and adherent to the human rights principles of non-discrimination and equality. Humans are exposed to climate change directly through changing weather patterns and indirectly through changes in water, air, food quality and quantity, ecosystems, agriculture, and economies.

Populations at greater risk from food insecurity, including smallholder and subsistence farmers, pastoralists, traditional societies, indigenous people, coastal populations, and artisanal fisherfolk, will suffer complex, localized impacts of climate change. These groups, whose adaptive capacity is constrained, will experience the negative effects on yields of low-latitude crops combined with a high vulnerability to extreme events. Indigenous people who rely on their natural resources for the provision of traditional foods will be especially severely affected (ACIA 2005; Kuhnlein 2003; Kuhnlein et al. 2002).

Climate change between 1970 and 2000 is estimated to have caused at least 160,000 deaths and 5 million disability-adjusted life years from malaria, diarrhea, malnutrition, and flooding (McMichael et al. 2004). Projected climate-change-related exposures are likely to affect the health status of millions of people, particularly those with low adaptive capacity, through such factors as:

- increased deaths, disease, and injury from heat-waves, floods, storms, fires, and droughts;
- increases in malnutrition and consequent disorders;
- altered spatial distribution of some infectious disease vectors; and
- increased burden of diarrheal diseases.

Most of the projected climate-related disease burden will result from increases in diarrheal diseases and malnutrition. Associations between monthly temperature and diarrheal episodes and between extreme rainfall events and monthly reports of outbreaks of water-borne disease have been reported worldwide (Checkley et al. 2000). Climate change is projected to increase the burden of diarrheal diseases in low-income regions by approximately 2 to 5 percent in 2020 and will impact low-income populations already experiencing a large burden of disease (Campbell-Lendrum et al. 2003; McMichael et al. 2004).

Gender Vulnerability

Men and women are affected differently in all phases of climate-related extreme weather events, from exposure to risk and risk perception; to preparedness behaviour, warning communication, and response; physical, psychological, social, and economic impacts; emergency response; and ultimately to recovery and reconstruction (Fothergill 1998). Many of the world's poorest people are rural women in developing countries who depend on subsistence agriculture to feed their families (Lambrou and Piana 2006). Climate change may also add to water and food insecurity and increase the labour burdens of women living in rural areas and developing countries, particularly in Africa and Asia (Parikh and Denton 2002).

Impacts on Food and Water Security

Climate change will affect all four dimensions of food security: food availability, stability of food supplies, access to food, and food utilization (FAO 2003a).

Food security depends not only on climate and socio-economic impacts but also, and crucially so, on changes to trade flows, stocks, and food-aid policy.

Food Availability

Agricultural output in developing countries is expected to decline by 10–20 percent by 2080, depending on whether there are beneficial effects from CO_2 "fertilization" (Cline 2007). Climate change and variability impacts on food production will be mixed and vary regionally (FAO 2003b), and will greatly exacerbate inequality in access to food. Recurrent severe droughts in several countries in Africa over the past three decades illustrate the potentially large effects of local and/or regional climate variability on crops and livestock (Hitz and Smith 2004; Fischer et al. 2005; Parry et al. 2005).

Evidence from models from the fourth IPCC assessment suggests that moderate local increases in temperature (1–3°C), along with the associated CO_2 increase and rainfall changes, can have small beneficial impacts on major rainfed crops (maize, wheat, rice) and pastures in mid- to high-latitude regions. In seasonally dry and tropical regions, even slight warming (1–2°C) reduces yield. Further warming (above a range of 1–3°C) has increasingly negative impacts on global food production in all regions. Temperature increases of more than 3°C may cause food prices to increase by up to 40 percent (Easterling et al. 2007).

Increases in temperature are leading to changes in the distribution of marine fisheries and community interactions (Parry et al. 2005). Regional changes in the distribution and productivity of particular fish species, as well as local extinctions, are expected due to continued warming (Easterling et al. 2007). Increases in atmospheric CO_2 are raising ocean acidity (The Royal Society 2005), which affects calcification processes, coral reefs' bleaching, and the balance of the food web. In relation to aquaculture production, increases in seawater temperature have been associated with increased densities of *Vibrio* spp in shellfish and harmful algal blooms, which are important causes of diarrhea and seafood toxicity respectively.

Global warming will confound the impact of natural variation on fishing activity and complicate management. The sustainability of the fishing industries of many countries will depend on increasing flexibility in bilateral and multilateral fishing agreements coupled with international stock assessments and management plans (Easterling et al. 2007).

Trade in cereal crops, livestock, and forestry products is projected to increase in response to climate change, with increased dependence on food imports for most developing countries. Exports of temperate zone food products to tropical countries will rise, while the reverse direction is likely for forestry trade in the short term (ibid.). This heightens vulnerability in developing countries to both climate and world market prices.

Food Stability and Access

Changes in the patterns of extreme weather events will affect the stability of, and access to, food supplies. Recent modelling studies suggest that increasing frequency of crop loss due to these extreme events may overcome positive effects of moderate temperature increases. This change in frequency of extreme events is likely to disproportionately impact smallholder farmers and artisanal fishers (ibid.). Climate-related animal and plant pests and diseases and alien invasive aquatic species will reduce the availability of quantities of food, influence the stability of the production system, and decrease food access through reduction of income from animal production, reduction of yields of food and cash crops, lowered forest productivity, and changes in aquatic populations, as well as increased costs of control (FAO 2008a).

Health and Food Utilization

Climate change may affect health outcomes and food utilization with additional consequences for malnutrition. For example, populations in water-scarce regions are likely to face decreased water availability, particularly in the subtropics. Flooding and increased precipitation are likely to contribute to increased incidence of infectious and diarrheal diseases. The risk of emerging zoonoses— animal diseases that can be transmitted to humans—may increase due to changes in the survival of pathogens in the environment, changes in migration pathways, carriers and vectors, and changes in the natural ecosystems.

Climate change plays an important role in the spatial and temporal distribution of vector-borne diseases such as malaria. In some areas, the geographical range of these diseases will contract in the long term, due to the lack of the necessary humidity and water for mosquito breeding.[2] Elsewhere, however, the geographical range of malaria will expand and the transmission season may be changed. It is estimated that in Africa climate change will increase the number of person-months of exposure to malaria by 16–28 percent by 2100 (McMichael 2004). Malaria affects food availability as well as access to and utilization of food by humans and livestock.

Impact Pathways

The impacts of climate change on food and water security and safety and on nutrition are a great concern, particularly for developing countries. These changes will have a profound impact on the fulfilment of human rights, in particular on the right to water, which is closely linked to the right to food. By 2080, it is estimated that 1.1 to 3.2 billion people will be experiencing water scarcity; 200 to 600 million, hunger; and 2 to 7 million more per year, coastal flooding (Yohe et al. 2007).

There are many pathways through which global climate change and variability may impact food and water security and safety, and nutrition, including:

- increased frequency of extreme climatic events;
- sea-level rise and flooding of coastal lands, leading to salination and/or contamination of water and agricultural lands;
- impacts of temperature increase and water scarcity on plant or animal physiology;
- beneficial effects to crop production through CO_2 fertilization;
- influence on plant diseases and pest species and livestock diseases, including zoonosis, leading to crop and animal losses; and
- damage to forestry, livestock, fisheries, and aquaculture.

In addition, multiple socio-economic and environmental stresses, such as globalization, limited availability of water resources, loss of biodiversity, the HIV/AIDS pandemic, and social and armed conflicts are further increasing sensitivity to climate change and reducing resilience in the agricultural sector (FAO 2003a).

Access to safe water remains an extremely important global health issue. More than two billion people live in the dry regions of the world and suffer disproportionately from malnutrition, infant mortality, and diseases related to contaminated or insufficient water (WHO 2005).

The impacts of climate change on freshwater systems and their management are mainly due to observed and projected increases in temperature, sea level, and precipitation variability. Climate change is likely to exacerbate declining reliability of irrigation water supplies, leading to increased competition for water for industrial, household, agricultural, and ecosystem uses. In coastal areas, sea-level rise will extend areas of salination of groundwater, resulting in a decrease in freshwater availability (Kundzewicz et al. 2007).

Links to Malnutrition

Research and information on the links between climate-change-related food and water insecurity and malnutrition are necessary. There is also a need for methodologies to convert estimated losses in regional crop yields into estimates of changes in numbers of malnourished people. This has been recognized as one of the critical research needs by the fourth IPCC assessment report.

Drought and water scarcity can lead to negative effects on nutrition through increased infections and mortality, and reduced food availability (in terms of both quantity and quality). During the 2000 drought in Gujarat, India, for instance, diets were found to be deficient in energy and several vitamins, and serious effects of drought on anthropometric indices may have been prevented by public-health measures (Hari Kumar et al. 2005). The HIV/AIDS epidemic may have further amplified the effect of drought on nutrition in countries such as those in Southern Africa (Mason et al. 2005). On the other hand, malnutrition increases the risk of both acquiring and dying from an infectious disease. For example, in Bangladesh both the impacts of drought and lack of food are associated with an increased risk of mortality from a diarrheal illness (Aziz et al. 1990).

Children in poor rural and urban slum areas are at high risk of diarrheal disease mortality and morbidity. Childhood mortality due to diarrhea in low-income countries, especially in sub-Saharan Africa, remains high, and child malnutrition is projected to persist in parts of low-income countries. Children who survive the acute illness may later die due to persistent diarrhea or malnutrition.

Climate Change and Sustainable Development

Sustainable development can reduce vulnerability to climate change by enhancing adaptive capacity and increasing resilience. On the other hand, climate change can slow the pace of progress toward sustainable development, either directly through increased exposure to adverse impact, or indirectly through erosion of the capacity to adapt (Yohe et al. 2007). Degradation of ecosystem services poses a barrier to achieving sustainable development and to meeting the MDGs (Millennium Ecosystem Assessment 2005).

In order to meet the MDGs, it would be necessary to balance competition for land for agriculture, livestock, forestry, and biofuels production. The expansion of livestock and biofuel sectors has a major role in deforestation and land degradation, and thereby contributes to climate change.

"Livestock's Long Shadow"

FAO's Livestock, Environment, and Development (LEAD) Initiative has identified the livestock sector as a major contributor to climate change, responsible for 18 percent of GHG emissions measured in CO_2 equivalent. The livestock sector is a key player in increasing water use, accounting for over 8 percent of global human water use, mostly for the irrigation of feedcrops. It is probably the largest sectoral source of water pollution and is the major driver of deforestation, as well as one of the leading drivers of land degradation, pollution, sedimentation of coastal areas, and facilitation of invasions by alien species (LEAD 2006).

There are measures that can help reduce the overall impact of livestock production. Among them, sustainable intensification can reduce effects on deforestation, pasture degradation, wildlife biodiversity, and resource use (Delgado et al. 1999). Emissions can be reduced through improved diets to reduce fermentation in ruminants' digestive systems and improved manure and biogas management. Water pollution and land degradation can be tackled through better irrigation systems, better management of waste, and improved diets that increase nutrient absorption.

Social Impacts of Climate Change

Implications for Rural and Urban Populations

Smallholder and subsistence-farming households in the dryland tropics are particularly vulnerable to increasing frequency and severity of droughts. These may lead to a higher likelihood of crop failure, increased diseases and mortality of livestock, indebtedness, out-migration, and dependency on food relief, with impacts on human development indicators such as health, nutrition, and education (Easterling et al. 2007).

Drought and the consequent loss of livelihoods is also a major trigger for migratory movements, particularly rural to urban migration. Population displacement to urban slums can lead to increases in diarrheal and other communicable diseases and poor nutritional status resulting from overcrowding and a lack of safe water, food, and shelter. Rural to urban migration contributes to the spread of HIV/AIDS, malaria, dengue fever, and other diseases (Confulonici et al. 2007).

Environmental Refugees and Social Conflict

The UN projects that there will be up to 50 million people escaping the effects of environmental deterioration by 2020. The spectrum of associated health risks includes food and water emergencies and infectious, nutritional, and mental diseases. By increasing the scarcity of basic food and water resources, environmental degradation increases the likelihood of violent conflict (LEAD 2006; Biggs et al. 2004). Conflict could emerge as a result of climate-change-related environmentally induced migration. Political refugees from violent regions are more likely to become involved in militant activities (Gleditsch, Nordås, and Salehyan 2007).

In sub-Saharan Africa, where cropping and grazing are often practised by different ethnic groups, the advance of crops into pasture land often results in conflict, as shown by major disturbances in the Senegal river basin between Mauritania and Senegal and in Northeast Kenya, between the Boran and the Somalis (Nori, Switzer, and Crawford 2005). According to the United Nations Environment Programme (2007), the conflict in Darfur has been driven in part by environmental degradation and exacerbated by climate change, and these forces threaten to trigger a succession of new wars across Africa.

Adaptation Strategies

In response to climate change, food-security-related adaptation strategies may be either autonomous or planned (Easterling et al. 2007). Autonomous adaptation is the ongoing implementation of existing knowledge and technology in response to the changes in climate experienced. Planned adaptation is the increase in adaptive capacity by mobilizing institutions and policies to establish or strengthen conditions favourable for effective adaptation and investment in new technologies and infrastructure.

Many of the autonomous adaptation options are extensions or intensifications of existing risk-management or production-enhancement activities for cropping systems, livestock, forestry, and fisheries production (Easterling et al. 2007). Autonomous adaptation strategies often have limitations. For example, native livestock breeds that are more heat tolerant often have lower levels of productivity.

Planned adaptation strategies can involve activities such as developing infrastructure or building the capacity to adapt in the broader user community and institutions, often by changing the decision-making environment under which autonomous adaptation activities occur (Easterling et al. 2007). Policy-based adaptations to climate change may include policies on natural resource management, human and animal health, governance, and political rights, among many others (Yohe et al. 2007).

If widely adopted, autonomous and planned adaptation strategies have substantial potential to offset negative climate change impacts and take advantage of positive ones (Easterling et al. 2007).

Mitigation Strategies

Agriculture, land use, and waste account for some 35 percent of the GHG emissions that contribute to climate change (Stern 2006). At the same time, improved agricultural practices can make a significant contribution at low cost to increasing soil carbon sinks and to GHG emission reductions. Key mitigation strategies in the agriculture sector include improved crop and grazing land management to increase soil carbon sequestration, restoration of degraded lands, improved rice cultivation, livestock and manure management to reduce methane emissions, and improved nitrogen fertilizer management to reduce nitrous oxide emissions (Metz et al. 2007).[3]

Improved management of tropical land offers a promising agriculture-based mitigation strategy. Reduced deforestation, more sustainable forest management, and adoption of agroforestry (integration of tree and crop cultivation) have particularly good potential to capture significant amounts of carbon and other GHGs and, at the same time, to contribute to poverty reduction. Agroforestry not only captures carbon and helps maintain soil health through nitrogen fixation and use of cuttings as fertilizer and mulch, but it also provides fodder, fruit, timber, fuel, medicines, and resins (CGIAR 2008).

Agricultural research can help create new technologies that will facilitate agriculture-based mitigation strategies. For example, research is underway at international agricultural research centres supported by the Consultative Group on International Agriculture Research to breed new, drought-tolerant varieties of sorghum, which will provide food, feed, and fuel from a single plant without current tradeoffs among uses (ibid.).

Priorities and Approaches for Responding to Threats to Nutrition from Climate Change

A combination of adaptation and mitigation measures, sustainable development, and research to enhance both adaptation and mitigation can diminish the threats to nutrition from climate change. Strategies should include measures that would simultaneously reduce pressures on biodiversity and food security and contribute to carbon sequestration. Such strategies can also contribute to making food prices more stable. The human rights framework offers the means to explicitly link environmental concerns to good governance and the inherent emphasis of human rights on "humans."

There are multiple adaptation options that imply different costs, ranging from changing practices to changing locations of food, fibre, forestry, and fishery activities. Changes in policies and institutions will be needed to facilitate adaptation for food security to climate change. On average, cereal cropping system adaptations such as changing varieties and planting times enable avoidance of a 10–15 percent reduction in yield corresponding to a 1–2°C local temperature increase (Easterling et al. 2007). The benefits of adaptation tend to increase with the degree of climate change up to a point; adaptive capacity in low latitudes is exceeded at 3°C local temperature increase (ibid).

With regard to mitigation, financial incentives can help promote improved land management, maintenance of soil carbon content, and efficient use of fertilizers and irrigation. This could reduce vulnerability to climate change, promote sustainable development, and help improve the health environment (Metz et al. 2007).

Adaptation and mitigation measures should be developed as part of overall and country-specific development programs such as Poverty Reduction Strategy Papers, pro-poor strategies, and national Food and Nutrition Action

Plans, with the engagement of all relevant stakeholders. Measures to reduce vulnerability should be included in disaster risk reduction plans. Donor agencies should assist developing countries to assess their capacity-building needs in this regard.

Agriculture, food, and nutrition issues need to be placed on national and international climate change agendas, in order to devise effective and pro-poor policies. The expiration of the Kyoto Protocol in 2012 offers an opportunity to bring these issues to the table as a new agreement is negotiated.

Adopting a human rights perspective when tackling the challenge of climate change puts people at the centre of attention of decision making. Sustaining and protecting the environment against degradation will be enhanced through the protection and promotion of human rights. At the same time, human rights cannot be fully realized without securing the environmental dimensions of ecosystem services essential to the right to life, the right to food, and all other human rights.

Notes

This chapter is adapted with permission from M.J. Cohen, C. Tirado, N.-L. Aberman, and B. Thompson (2008), "Impact of Climate Change and Bioenergy on Nutrition" (Washington and Rome: International Food Policy Research Institute and Food and Agriculture Organization of the United Nations), http://www.ifpri.org/pubs/cp/cohen2008climate/cohenetal2008climate.pdf or http://www.fao.org/docrep/010/ai799e/ai799e00.htm.

1   The IPCC is a scientific intergovernmental body set up by the UN World Meteorological Organization and the UN Environment Programme to provide decision makers and others with objective information about climate change. The scientific community generally regards its reports as authoritative (see Sample 2007).

2   The northern limit of *Plasmodium falciparum* malaria in Africa is the Sahel, where rainfall is an important limiting factor in disease transmission.

3   Nitrogen fertilizer tends to break down into nitrous oxide, a greenhouse gas that also contributes to ozone depletion, and nitrate, which aids crop growth but also contaminates streams and groundwater, thereby threatening health and nutrition.

Works Cited

Arctic Climate Impact Assessment (ACIA) (2005). *Arctic Climate Impact Assessment.* New York: Cambridge University Press.

Aziz, K.M.A., B.A. Hoque, S. Huttly, K.M. Minnatullah, Z. Hasan, M.K. Patwary, M.M. Rahaman, and S. Cairncross (1990). "Water Supply, Sanitation and Hygiene Education: Report of a Health Impact Study in Mirzapur, Bangladesh." Water and Sanitation Report Series No. 1. Washington, DC: World Bank.

Boko, M., I. Niang, A. Nyong, C. Vogel, A. Githeko, M. Medany, B. Osman-Elasha, R. Tabo and P. Yanda (2007). "Africa." *Climate Change 2007: Impacts, Adaptation and Vulnerability. Contribution of Working Group II to the Fourth Assessment Report of the Intergovernmental Panel on Climate Change*, ed. M.L. Parry, O.F. Canziani, J.P. Palutikof, P.J. van der Linden and C.E. Hanson, 433–67, Cambridge University Press, Cambridge, UK.

Biggs, R., E. Bohensky, P.V. Desanker, C. Fabricius, T. Lynam, A. Misselhorn, C. Musvoto, M. Mutale, B. Reyers, R.J. Scholes, S. Shikongo, and A.S. van Jaarsveld (2004). "Nature Supporting People: The Southern Africa Millennium Ecosystem Assessment." Pretoria: Council for Scientific and Industrial Research.

Campbell-Lendrum, D., A. Pruss-Ustun, and C. Corvalan (2003). "How Much Disease Could Climate Change Cause?" In *Climate Change and Human Health: Risks and Responses*, ed. A. McMichael, D. Campbell-Lendrum, C. Corvalan, K. Ebi, A. Githeko, J. Scheraga, and A. Woodward, 133–59, Geneva: World Health Organization/World Meteorological Organization/UN Environment Programme.

Checkley, W., L.D. Epstein, R.H. Gilman, D. Figueroa, R.I. Cama, J.A. Patz, and R.E. Black (2000). "Effects of El Niño and Ambient Temperature on Hospital Admissions for Diarrhoeal Diseases in Peruvian Children." *The Lancet* 355, no. 1902: 442–50.

Cline, W.R. (2007). *Global Warming and Agriculture: Impact Estimates by Country.* Washington, DC: Center for Global Development and Peterson Institute for International Economics.

Cohen, M.J., C. Tirado, N.-L. Aberman, and B. Thompson (2008). "Impact of Climate Change and Bioenergy on Nutrition." Washington and Rome: International Food Policy Research Institute and Food and Agriculture Organization of the United Nations. http://www.ifpri.org/PUBS/cp/cohen2008climate/cohenetal2008climate.pdf or http://www.fao.org/docrep/010/ai799e/ai799e00.htm.

Confalonieri, U., B. Menne, R. Akhtar, K.L. Ebi, M. Hauengue, R.S. Kovats, B. Revich and A. Woodward (2007). "Human Health." *Climate Change 2007: Impacts, Adaptation and Vulnerability. Contribution of Working Group II to the Fourth Assessment Report of the Intergovernmental Panel on Climate Change*, ed. M.L. Parry, O.F. Canziani, J.P. Palutikof, P.J. van der Linden and C.E. Hanson, 391–431, Cambridge University Press, Cambridge, UK.

Consultative Group on International Agricultural Research (CGIAR) (2008). "Global Climate Change: Can Agriculture Cope?" Washington, DC. http://www.cgiar.org/impact/global/climate.html.

Cruz, R.V., H. Harasawa, M. Lal, S. Wu, Y. Anokhin, B. Punsalmaa, Y. Honda, M. Jafari, C. Li and N. Huu Ninh (2007). "Asia." *Climate Change 2007: Impacts, Adaptation and Vulnerability. Contribution of Working Group II to the Fourth Assessment Report of the Intergovernmental Panel on Climate Change*, ed. M.L. Parry, O.F. Canziani, J.P. Palutikof, P.J. van der Linden and C.E. Hanson, 469–506, Cambridge University Press, Cambridge, UK.

Delgado, C., M. Rosegrant, H. Steinfeld, S. Ehui and C. Courbois (1999). "Livestock to 2020: The Next Food Revolution." 2020 Vision for Food, Agriculture, and the Environment Discussion Paper No. 28. Washington, DC: International Food Policy Research Institute.

Easterling, W.E., P.K. Aggarwal, P. Batima, K.M. Brander, L. Erda, S.M. Howden, A. Kirilenko, J. Morton, J.-F. Soussana, J. Schmidhuber and F.N. Tubiello (2007). "Food, fibre and forest products." *Climate Change 2007: Impacts, Adaptation and Vulnerability. Contribution of Working Group II to the Fourth Assessment Report of the Intergovernmental Panel on Climate Change*, ed. M.L. Parry, O.F. Canziani, J.P. Palutikof, P.J. van der Linden and C.E. Hanson, 273-313, Cambridge University Press, Cambridge, UK.

Food and Agriculture Organization of the United Nations (2003a). "Strengthening Coherence in FAO's Initiatives to Fight Hunger." Conference, Thirty-second Session. 29 November to 10 December. Rome.

——— (2003b). "Impact of Climate Change on Food Security and Implications for Sustainable Food Production." Committee on World Food Security. Conference, Twenty-ninth Session. 12–16 May. Rome.

——— (2008a). "Expert Meeting on Climate-Related Transboundary Pests and Diseases Including Relevant Aquatic Species, Food and Agriculture Organization of the United Nations, 25–27 February 2008, Options for Decision Makers." Rome. http://www.fao.org/fileadmin/user_upload/foodclimate/presentations/diseases/Options EM3.pdf.

——— (2008b). "The State of Food Insecurity in the World 2008." Rome. ftp://ftp.fao.org/docrep/fao/011/i0291e/i0291e00.pdf.

Fischer, G., M. Shah, F.N. Tubiello, and H. Van Velthuizen (2005). "Integrated Assessment of Global Crop Production." *Philosophical Transactions of the Royal Society B*. 360, no. 1463: 2067–83.

Fothergill, A. (1998). "The Neglect of Gender in Disaster Work: An Overview of the Literature." In *The Gendered Terrain of Disaster: Through Women's Eyes*, ed. E. Enarson and B. Morrow. Westport, CT: Praeger.

Gleditsch, N.P., Ragnhild Nordås, and Idean Salehyan (2007). "Climate Change and Conflict: The Migration Link." Coping with Crisis Working Paper. New York: International Peace Institute. http://www.ipacademy.org/asset/file/169/CWC_ Working_Paper_Climate_Change.pdf.

Hari Kumar, R., K. Venkaiah, N. Arlappa, S. Kumar, G. Brahmam, and K. Vijayaraghavan (2005). "Diet and Nutritional Status of the Population in the Severely Drought Affected Areas of Gujarat." *Journal of Human Ecology* 18, no. 4: 319–26.

Hitz, S., and J. Smith (2004). "Estimating Global Impacts from Climate Change." *Global Environmental Change* 14, no. 3: 201–18.

Kuhnlein, H.V. (2003). "Micronutrient Nutrition and Traditional Food Systems of Indigenous Peoples." *Food, Nutrition and Agriculture* 32. Rome: Food and Agriculture Organization.

Kuhnlein, H.V., H.M. Chan, D. Leggee, and V. Barthet (2002). "Macronutrient, Mineral and Fatty Acid Composition of Canadian Arctic Traditional Food." *Journal of Food Composition and Analysis* 15, no. 5: 545–66.

Kundzewicz, Z.W., L.J. Mata, N.W. Arnell, P. Döll, P. Kabat, B. Jiménez, K.A. Miller, T. Oki, Z. Sen and I.A. Shiklomanov (2007). "Freshwater resources and their management." *Climate Change 2007: Impacts, Adaptation and Vulnerability. Contribution of Working Group II to the Fourth Assessment Report of the Intergovernmental Panel on Climate Change,* ed. M.L. Parry, O.F. Canziani, J.P. Palutikof, P.J. van der Linden and C.E. Hanson, 173–210, Cambridge University Press, Cambridge, UK.

Lambrou, Y., and G. Piana (2006). "Energy and Gender Issues in Rural Sustainable Development." Rome: Food and Agriculture Organization. ftp://ftp.fao.org/ docrep/fao/010/ai021e/ai021e00.pdf.

Livestock, Environment and Development (LEAD) (2006). "Livestock's Long Shadow: Environmental Issues and Options." Rome: Food and Agriculture Organization. http://www.fao.org/docrep/010/a0701e/a0701e00.htm.

Lemke, P., J. Ren, R.B. Alley, I. Allison, J. Carrasco, G. Flato, Y. Fujii, G. Kaser, P. Mote, R.H. Thomas, and T. Zhang (2007). "Observations: Changes in Snow, Ice and Frozen Ground." In Solomon et al. 2007.

Mason, J.B., A. Bailes, K.E. Mason, O. Yambi, U. Jonsson, C. Hudspeth, P. Hailey, A. Kendle, D. Brunet, and P. Martel (2005). "AIDS, Drought and Child Malnutrition in Southern Africa." *Public Health Nutrition* 8, no. 6: 551–63.

McMichael, A., D. Campbell-Lendrum, S. Kovats, S. Edwards, P. Wilkinson, T. Wilson, R. Nicholls, S. Hales, F. Tanser, D. Le Sueur, M. Schlesinger, and N. Andronova (2004). "Global Climate Change." In *Comparative Quantification of Health Risks: Global and Regional Burden of Disease due to Selected Major Risk Factors,* vol. 2, ed. M. Ezzati, A. Lopez, A. Rodgers, and C. Murray, 1543–1649, Geneva: World Health Organization.

Meehl, G.A., T.F. Stocker, W.D. Collins, P. Friedlingstein, A.T. Gaye, J.M. Gregory, A. Kitoh, R. Knutti, J.M. Murphy, A. Noda, S.C.B. Raper, I.G. Watterson, A.J. Weaver, and Z.-C. Zhao (2007). "Global Climate Projections." In Solomon et al. 2007.

Metz, B., O.R. Davidson, P.R. Bosch, R. Dave, and L.A. Meyer, eds. (2007). *Climate Change 2007: Mitigation. Contribution of Working Group III to the Fourth Assessment Report of the Intergovernmental Panel on Climate Change.* Cambridge: Cambridge University Press.

Millennium Ecosystem Assessment (2005). *Ecosystems and Human Well-being: Synthesis.* Washington, DC: Island Press.

Nicholls, R.J., P.P. Wong, V.R. Burkett, J.O. Codignotto, J.E. Hay, R.F. McLean, S. Ragoonaden, and C.D. Woodroffe (2007). "Coastal Systems und Low-lying Areas." *Climate Change 2007: Impacts, Adaptation and Vulnerability. Contribution of Working Group II to the Fourth Assessment Report of the Intergovernmental Panel on Climate Change,* ed. M.L. Parry, O.F. Canziani, J.P. Palutikof, P.J. van der Linden and C.E. Hanson, 173–210, Cambridge University Press, Cambridge, UK.

Nori, M., J. Switzer, and A. Crawford (2005). "Herding on the Brink: Towards a Global Survey of Pastoral Communities and Conflict." Winnipeg: International Institute for Sustainable Development. http://www.iisd.org/pdf/2005/security_herding_on_brink.pdf.

Parikh, J. K., and F. Denton (2002). "Gender and Climate Change." *Tiempo* 47. March. http://www.cru.uea.ac.uk/tiempo/floor0/recent/issue47/t47a7.htm.

Parry, M.L., C. Rosenzweig, and M. Livermore (2005). "Climate Change, Global Food Supply and Risk of Hunger." *Philosophical Transactions of the Royal Society B.* 360, no. 1463: 2125–38.

Royal Society, The (2005). *Ocean Acidification Due to Increasing Atmospheric Carbon Dioxide.* Cardiff: Clyvedon Press.

Sample, Ian (2007). "Scientists Offered Cash to Dispute Climate Study." *The Guardian.* 2 February.

Solomon, S., D. Qin, M. Manning, Z. Chen, M. Marquis, K.B. Averyt, M. Tignor, and H.L. Miller, eds. (2007). *Climate Change 2007: The Physical Science Basis, Contribution of Working Group I to the Fourth Assessment Report of the Intergovernmental Panel on Climate Change.* Cambridge: Cambridge University Press.

Stern, Nicholas (2006). *Stern Review on the Economics of Climate Change.* Cambridge: Cambridge University Press.

Trenberth, K.E., P.D. Jones, P. Ambenje, R. Bojariu, D. Easterling, A. Klein Tank, D. Parker, F. Rahimzadeh, J.A. Renwick, M. Rusticucci, B. Soden, and P. Zhai (2007). "Observations: Surface and Atmospheric Climate Change." In Solomon et al. 2007.

United Nations Environment Programme (2007). "Sudan Post-Conflict Environmental Assessment." Nairobi. http://postconflict.unep.ch/publications/UNEP_Sudan.pdf.

World Health Organization (2005). *Ecosystems and Human Wellbeing: Health Synthesis, A Report of the Millennium Ecosystem Assessment.* Geneva. http://www.millenniumassessment.org/documents/document.357.aspx.pdf.

Yohe, G.W., R.D. Lasco, Q.K. Ahmad, N.W. Arnell, S.J. Cohen, C. Hope, A.C. Janetos, and R.T. Perez. 2007. "Perspectives on climate change and sustainability." *Climate Change 2007: Impacts, Adaptation and Vulnerability. Contribution of Working Group II to the Fourth Assessment Report of the Intergovernmental Panel on Climate Change*, ed. M.L. Parry, O.F. Canziani, J.P. Palutikof, P.J. van der Linden, and C.E. Hanson, 811–41, Cambridge University Press, Cambridge, UK.

CHAPTER 11

Fossil Energy and the Biophysical Roots of the Food Crisis

Tony Weis

Champions of industrial agriculture celebrate long-term yield and productivity gains together with steadily declining prices. These are, on one hand, basic quantifiable facts and have underpinned foundational assumptions about modern societies, such as the beliefs that development entails progressively reducing the share of agricultural labour in the workforce and that improving diets means moving up the protein ladder toward more meat intensive diets. But, as this chapter argues, the cheap bounty of industrial agriculture might also be seen to constitute a profound and dangerous illusion, one that reflects a perverse system of valuation and cost-accounting that has long partially obscured its unsustainability. Further, it suggests that such an illusion is bound to shatter under the weight of intensifying biophysical instabilities, with the dramatic food price volatility in world markets representing initial cracks in a speeding systemic crisis.

In order to appreciate this crisis stage, and the uneven vulnerability to the associated price increases and volatility, it is necessary to begin by exploring the nature of industrial agriculture, and in particular the substitution of labour with technology and its dependence upon on fossil energy and derivatives. This, in turn, provides the basis for understanding the most proximate trigger in the shift from chronic instability to increasingly acute systemic crisis: the looming scarcity of fossil energy supplies, or "peak oil," which is bound to be reflected in the rising costs of industrial methods and long-distance flows of inputs and outputs.

These cost pressures, however, pale before the still largely unaccounted costs of inaction on climate change, now commonly referred to as the "business-as-usual" scenario. Industrial agriculture is a major source of greenhouse gas emissions at the same time as the failure to make large and rapid emission reductions threatens to undermine a crucial aspect of all agriculture, the relative climatic stability of the Holocene (the last 10,000 years during which time human civilization arose). Yet because the magnitude of future costs are still not significantly measured in economic terms, they have not destabilized the operative logic of industrial agriculture or its dominant actors, agro-transnational corporations (TNCs).

Rather, in the face of peak oil and climate change, industrial agriculture is being framed both as a "technological fix" for the looming scarcity of liquid fuel and as a "green" source of energy, a momentous new dynamic in global agricultural production that is critically reviewed here. In spite of very dubious energy budgets, the biofuel boom has ironically buoyed industrial farmers and agro-TNCs in the short-term. This has created a strong pull on industrial grain and oilseed supplies that has magnified the price pressures associated with the continuing growth in demand for livestock feed as meat consumption expands.

A primary objective of this chapter is to assess the destructive market signals guiding industrial agriculture in the face of worsening biophysical instabilities. Although the potentially catastrophic costs of climate change inaction continue to go largely unregistered (at least at a scale where they might drive serious action to mitigate the magnitude of climate change), the scarcity of a fundamental resource is beginning to register—still with considerable volatility as an input cost, but also as a basis of enormous market potential and profits for industrial farmers and agro-TNCs. Recognizing this systemic illogic helps to make sense of recent food price volatility, the responses of dominant actors, the highly regressive social fallout, and the great danger that, in the absence of major political and economic changes, much worse is ahead. However, the more this destructive course is understood, the greater the chance that more democratic, socially just, and ecologically rational transitions might be initiated. The chapter concludes in the spirit that systemic crises always contain opportunities for change.

Imbalances and Instability

Roughly half the world's agricultural exports—and the majority of all grain, oilseed, and livestock exports—come from a very narrow base of large-scale industrialized producers in countries such as the United States, Canada, Brazil, Argentina, France, and Australia. The flipside of this is the precarious net food-import dependence of many of the world's poorest countries, which typically have the largest agrarian populations, in a world where nearly one billion people are malnourished (Weis 2007; Rosset 2006). The widespread rioting that accompanied recent food price rises puts the social instability associated with this imbalance in vivid display.

Agro-TNCs control much of the value and decision-making on both the input and output sides of industrial agriculture, as well as in global agricultural trade. The agro-input industry is an oligopolistic web in which enormous corporations control increasing shares of the related and overlapping global markets for chemicals, seeds, fertilizers, and animal pharmaceuticals. Corporate control over processing, distribution, and retailing has also intensified and squeezed value out of agricultural production.

Temperate agricultural systems are dominated by the industrial grain-oilseed-livestock complex, with the "big three" cereals (maize, rice, and wheat) and the dominant oilseed (soybeans) providing half of the world's plant-based calories. The coupling of mechanization, enhanced seeds, increased irrigation, and intensified fertilizer and chemical inputs brought dramatic yield gains in the second half of the twentieth century that, on a global scale, drove a roughly 50 percent increase in per capita grain production and a 100 percent rise in per capita meat production during a period of rapid human population growth (Weis 2007).

Industrial maize and soybean production have underpinned the highly uneven growth in per capita meat consumption, the intensification of farm animal rearing, and the speeding "turnover time" of animals. The world livestock "inventory" (i.e., the population at any given moment) tops 20 billion animals, and more than 60 billion animals are slaughtered annually. The "big three" livestock species (pigs, chickens, and cattle) account for almost 90 percent of all animal flesh produced in the world, with large and growing populations of chickens and pigs reared in factory farm conditions that depend upon concentrated feed and chemical and pharmaceutical inputs. Per capita meat consumption and the population of animals reared in factory conditions continue to grow relentlessly, now stoked by fast-rising demand in Asia as well as in other industrializing countries like Brazil (Weis 2007; Nierenberg 2005).

Economies of scale in industrial agriculture have been associated with the profound polarization of landholding and productivity, as farmers trapped in a long-term "cost-price squeeze" (i.e. rising input costs and falling farm-gate prices) have been forced to "get big or get out." This trajectory has been further fortified both by explicit subsidies—particularly in the United States and European Union (EU)—and by the implicit subsidization contained in a vast range of environmental externalities: high rates of soil erosion; persistent toxicity; the overdraft of rivers, streams, and underground water; the salinization of over-irrigated soils; air and water pollution associated with factory farming; disease threats like avian flu, listeriosis, and mad cow; large greenhouse gas emissions; and the immeasurable suffering of soaring populations of sentient beings. Finally, given the immense historic and contemporary tensions entwined in struggles to control the world's oil supply, industrial agriculture could also conceivably be seen to involve a "geopolitical externality." That is, in addition to the unaccounted costs of carbon emissions, the relatively cheap price of oil has long failed to reflect the vast expenditure on military infrastructure and incursions surrounding the world's oil reserves, most notably the hundreds of US bases in the Middle East (Foster et al. 2008).

Taken together, mechanized scale, input-augmented yields, and explicit and implicit subsidies led industrial food prices to decline in real terms over much of the past half-century. From 1960 to 2000, the real prices of the big three cereals declined by 60 percent, and from 1974 to 2005, the decline in world market prices for a total food index was even greater, falling by an astonishing 75 percent (*The Economist* 2007; FAO 2002). As cheap, industrial surpluses were projected through aid, dumping, and commercial trade, they put downward pressure on food prices in many labour-intensive agricultural systems, impoverishing smallholder agriculture and fostering the deep food-import dependencies in many low-income countries noted above (Weis 2007; Rosset 2006; FAO 2002).[1]

This low-price bounty has long been ascribed to competitiveness, efficiency, and the inevitable triumph of industrial agriculture and is celebrated in leading development theories for displacing—or, in rosier terms, "releasing"—small farmers into more modern livelihoods. But such theories have ignored how the efficiency and competitiveness of industrial agriculture has been braced by a large range of unaccounted costs and how the substitution of agrarian labour with technology hinges on fossil energy and derivatives.

Fossil Energy and the Substitution of Agrarian Labour

One of the most elemental tendencies of industrial capitalism is to substitute labour with technology wherever possible. This is at the heart of economies of scale and has long depended upon the extraction of fossilized biomass, stored solar energy from millions of years ago.

Agriculture poses particular problems for mechanization and economies of scale, as the standardization of plant life across large areas of a landscape creates a host of intractable biophysical problems with soils, insects, weeds, and diseases. Bare ground between planted rows and mechanized plowing, planting, and spraying increase susceptibility to soil erosion and nutrient loss, with compaction by heavy machinery creating further problems. Large-scale monocultures effectively mine the soil and cannot be sustained very long without external sources of nutrients given the rates of erosion and fertility degradation. By far the biggest source of replaced nutrients comes from synthetic nitrogen fertilizer, manufactured using natural gas, which has been inseparable from the yield gains of industrial agriculture.

Synthetic nitrogen fertilizer is a large source of energy consumption and carbon emissions, through its manufacture as well as in transport and application, given its bulky character and the diffuse nature of agriculture. After nitrogen, the next most important soil nutrients lost are phosphorous and potassium, which can also only be replaced from a non-renewable base for a finite period of time.[2]

Although champions of the Green Revolution like to point out how grain yields per hectare grew by a factor of 2.4 between 1950 and 1990, this was accompanied by a more than tenfold rise in synthetic fertilizer use (Brown 1996). Further, the Food and Agriculture Organization of the United Nations (FAO) notes that the rate of yield gains stemming from industrial inputs was much faster in the 1960s and 1970s than in the quarter-century since, which implies an effective slowing down of the Green Revolution amid a massive draw-down of the resources fuelling it.

In addition to causing soil problems, large-scale biological homogenization increases vulnerability to the rapid spread of pests, weeds, fungus, or disease, making monocultures dependent on a range of petrochemical-based herbicides, insecticides, and fungicides. This is magnified by the fact that the protracted use of chemicals tends to have a treadmill effect, as insect and weed resistance develops over time and natural predators and controls are eliminated. Further, the transportation fuel consumption associated with rising "food miles" (i.e., the long-distance transport of food from land to mouth) cannot be separated from the nature of on-farm production, because the substitution of labour with technology has abetted the increasing control of centralized corporate intermediaries between producer and consumer and hence is implicated in the separation of consumers from local agricultural landscapes.

As noted earlier, industrial grain and oilseed monocultures are closely linked to rising industrial livestock production and the "meatification" of human diets. World livestock populations have increased far beyond rangeland stocking capacities in proliferating factory farms. In cycling grains and oilseeds through livestock, large percentages of plant protein, carbohydrates, and fibre are lost, which means that agriculture gains a more expansive footprint in the landscape as the level of meat consumption grows—something which might be understood as our expanding "ecological hoofprint" (Weis 2007).

In addition to the increasing volumes of industrial grains and oilseeds (and the associated inputs) which growing livestock populations necessitate, factory farms, slaughterhouses, processing plants, and refrigerated shipping and retailing consume vast amounts of energy. Thus, much more energy goes into a unit of protein derived from factory-farmed meat than a unit of protein from grain (Nierenberg 2005). When this aggregated fossil energy consumption is coupled with the role of expanding livestock populations in global deforestation and in rising methane and nitrous oxide emissions, the result is a very large atmospheric burden. The Intergovernmental Panel on Climate Change (IPCC) has identified global livestock expansion as one of the leading causes of anthropogenic climate change (Black 2008).

In short, fossil energy and derivatives and greenhouse gas emissions are embedded in every calorie of cheap, industrial food, and in the "progress" toward ever more meat-intensive diets. Unfortunately, despite the urgency of large reductions, the atmospheric burden continues to go virtually unvalued in cheap industrial food. But the pressures associated with the growing scarcity of fossil energy and derivatives are inescapable and the most proximate reason that the deceptive efficiency of industrial agriculture is cracking.

Peak Oil, Industrial Agriculture, and the Biofuel Boom

The International Energy Agency (IEA 2007) estimates that fossil energy accounts for 80 percent of the world's total primary energy supply: oil 34 percent, coal 25 percent, and natural gas 21 percent. For the Organization for Economic Co-operation and Development (OECD) this is even greater—84 percent of total supply: oil 41 percent, natural gas 22 percent, and coal 21 percent.

But with production now far outpacing new discoveries, it is widely recognized that humanity is either fast approaching or has just passed the half-way point in the consumption of the earth's oil supply, a phenomenon referred to as peak oil. It is a certainty that the extraction of the back half of this supply will be increasingly difficult, and hence energy demanding and costly, in addition to more obvious supply–demand pressures (Heinberg 2005). As the oil minister of the United Arab Emirates recently put it, "the age of easy oil is gone forever" (*The Economist* 2008). There are still bound to be a few blips in world oil markets based on short-term supply and demand, as witnessed amid the financial turbulence of the world economy, but as one Canadian oil industry analyst explains, falling prices as the economy contracts are bound to eventually "morph into more lasting fears of supply destruction" (Hamilton 2008).

Despite inevitable supply constraints, demand for oil is projected to continue increasing in the short term. In 2007, global oil consumption was 84.6 million barrels per day (M bpd), and by 2030, the IEA projects this to be 102.3 M bpd. In a related projection, there are expected to be 450 million more vehicles on the road worldwide by 2030, roughly 50 percent more than at present. While there are obvious uncertainties, most estimates place remaining reserves in the range of just over one trillion barrels, roughly the same as what has been consumed since age of oil began in the late nineteenth century (BP Global 2008; *Oil & Gas Journal* 2006). There are also clear uncertainties about the rate of growth in consumption, but even holding 2007 consumption levels constant (probably a serious underestimation), global reserves will only extend a few decades, a reality starkly presented in a recent Chevron ad: "It took us 125 years to use the first trillion barrels of oil. We'll use the next trillion in 30."[3] As suggested at the outset of this chapter, the era of peak oil is pulling industrial agriculture in two basic and opposed ways: first, as a "fix" for the shortage of liquid fuel, and, second, as the biophysical overrides needed for mechanization and monocultures as these become more scarce and costly.

There are large and varied technological challenges associated with scaling-up nuclear, hydro, tidal, solar, wind, and other sources of energy to the substitute the major contribution that fossil fuels make within most energy grids, but there are a significant range of already well-established technologies, coupled with large gains to be made in conservation and efficiency of the built environment. Replacing liquid fuels is immensely more difficult and poses vexing questions given how the compression of time and space—and hence global trade, travel, and geostrategic power—hinge on oil, which accounts for virtually all liquid fuel consumed today. In this context, the desperate search for alternative liquid fuels is clear.

In principle, biofuels represent the possibility that the sun's energy might be renewably converted into liquid form via photosynthesis, with the additional promise of burning more cleanly. If the carbon accumulated in growing the biofuel plant source is simply released through the combustion of the biofuels, then depending on how it was processed it might even conceivably represent a carbon neutral energy source, as some have sought to portray it. Unfortunately, the benefits and costs of biofuel production are a lot more complex than they are portrayed by their emerging legion of corporate, political, auto, and big-farmer advocates, and it is a biophysical impossibility that biofuels might provide a large-scale replacement for oil.

Biofuels are typically classed in terms of "first" and "second" generations. First generation biofuels are essentially ethanol (predominantly derived from maize and sugar) and biodiesel (predominantly soy) and are commonly blended with traditional petroleum sources. Second generation biofuels are liquid fuels derived from certain non-edible grasses, woody biomass, straw, and some wastes (significantly, non-food portions of plants), with the key being the development of enzymes capable of converting plant cellulose into ethanol. The prospective benefit is that they could greatly improve input–yield ratios, enhance the conservation of carbon in soils because they can be made from permanent crops, and greatly expand the potential scale of source material while reducing competition with food supplies. However, the commercial viability of second-generation biofuels is still likely years away,[4] and even at the most ambitious targets they will only be able to substitute a fraction of current petroleum consumption (WorldWatch and CAP 2006).

Given the research still needed on second generation biofuels and the large fixed investments in processing ethanol, the biofuel boom appears centred squarely on first generation biofuels for the next decade at least and possibly beyond. Between 2000 and 2005, US production of ethanol doubled, and in 2006 it overtook Brazil's long-established industry (centred on sugar) as the world's largest producer of biofuels. In addition to the United States and Brazil, many other countries are racing into biofuels. From 2006 to 2007, the total world volume of coarse grains converted to biofuels increased by 15 percent, led by maize for ethanol. Both the EU (where soy for biodiesel is more common) and India have set goals of replacing at least 10 percent of vehicle fuel demand with biofuels, while China is beginning to devote more maize to ethanol (Sample 2007). Indonesia and Malaysia are converting large areas of tropical forest into palm oil plantations to produce biodiesel.

Yet there is growing evidence of the poor energy budgets of biofuels, when the fossil energy used in production (i.e., in the inputs, farm machinery, irrigation, processing, transportation, and fermentation and distillation) is weighed against the energy contained in biofuels. Research is showing that more fossil energy is going into most forms of biofuel production than is coming out as liquid fuel. Maize ethanol is the most glaring because it is so prominent and because maize is such a soil impoverishing crop and the leading consumer of pesticides and nitrogen fertilizer (Patzek and Pimentel 2006; Pimentel and Patzek 2005). Where energetic margins for first generation biofuels are not negative they are invariably thin, and even thin net positive margins for biofuels (i.e., more liquid energy comes out than fossil energy goes in) do not help mitigate against climate change.

An enormous amount of land must be converted to biofuels to make a small dent in current levels of fossil energy consumption. For instance, Righelato and Spracklen (2007) estimate that to substitute just 10 percent of current demand for gas and diesel with biofuels in the United States would require 43 percent of its total current cropland area. This entails rising pressure not only on food supplies and prices but also on forests and grasslands. To increase the footprint of agriculture and clear more forests and grasslands would entail significant volumes of carbon released in the immediate term, as carbon stored in the vegetation and soils is rapidly oxidized, and a declining capacity for carbon sequestration in the longer term—a double climate change whammy (the apex of which is the conversion of tropical rainforests to palm oil plantations). Thus, the expansion of first generation biofuels has a destabilizing net impact on carbon cycles, and, as Righelato and Spracklen argue, from the perspective of climate change mitigation, land devoted to biofuels would be better restored to forests to sequester carbon.

To understand the biofuel boom in light of its biophysical irrationality, Monbiot (2007) suggests that it be seen as a means of political avoidance. Instead of facing difficult questions about individual power over space in an age of climate change and peak oil, biofuels "create the impression that governments can cut carbon emissions and ... keep expanding the transport networks.... No one has to be confronted." This political objective has, in turn, fit nicely with the interests of agro-TNCs and large-scale, industrial farmers, summed up powerfully in a recent advertisement of the world's largest grain processor:

> The world's demand for energy will never stop growing.
> Which is why a farmer is growing corn.
> And a farmer is growing soy.
> And why ADM is turning these crops into biofuels.
> The world's demand for energy will never stop.
> Which is why ADM will not stop.
> We're only getting started.

Uneven Vulnerability to the Fading Promise of Cheap Food

> "If you combine the increase of the oil prices and the increase of food prices then you have the elements of a very serious crisis in the future."
> – Jacques Diouf, director-general of the FAO, October 2007

Though hundreds of millions remained hungry, increasing grain yields, more meat-intensive diets, and falling prices have long been pointed to as evidence of the robustness of corporate-dominated industrial agriculture. Now, the promise of cheap food is fast ending, even before the productive model breaks down, with converging biophysical problems of climate change, water shortages, and land degradation on the horizon. The intensity of demand pressures is already such that the dramatic price rises occurring from 2005 to 2007 came amid strong overall global production levels, and the volumes of grains and oilseeds devoted to fuel and feed are projected to continue growing in the years ahead.

As noted, many nations have set ambitious biofuel substitution targets, which Brown (2007) frames as an "epic" confrontation between the world's 800 million motorists and its 2 billion poorest people. Poor energy budgets coupled with the stark inequality of "feeding cars not people" has, not surprisingly, drawn rising indignation (Monbiot 2007). For instance, in 2007 the Special Rapporteur on the Right to Food of the UN Human Rights Council described biofuels as a "crime against humanity" (Ferrett 2007), while *The Economist* (2007) noted that filling "an SUV's fuel tank with ethanol...[requires] enough maize to feed a person for a year."

In addition to the biofuel boom, the "meatification" of diets—intensification and growing scale of livestock production and the associated pull of feedstock—is another major fossil-energy-related dimension in the pressures on grain and oilseed prices, and this is also projected to continue increasing in the coming decades. An estimate by the FAO suggested that global per capita meat consumption will grow a further 44 percent by 2030 (FAO, 2002). At the forefront of this dietary transformation is China, where per capita meat consumption has increased by 150 percent since 1980, with concerted state efforts planning further growth (*The Economist* 2007). Although China still has a relatively labour-intensive system of agriculture, it is industrializing rapidly, particularly with respect to livestock rearing, and this has enormous implications for both fossil energy consumption and greenhouse gas emissions. Fossil energy input into Chinese agriculture grew one-hundred-fold in the second half of the twentieth century alone (Goodland and Pimentel 2000).

The great supply-side stress in the sharp food price rises from 2005 to 2007 relates to a concurrent spike in fossil energy and input prices, recalling the large energy budget contained in the production and transport of industrial fertilizers and the long food miles that typify industrial food systems. The FAO (2008) highlights how an index of foodstuff prices has moved largely together with index of fertilizer and crude oil prices, following a similar trajectory but at slightly lower rates.

As these supply and demand pressures are layered onto already extensive global food insecurity, the most vulnerable are the poor in the world's low-income countries, where dependence upon global breadbasket regions has long been cultivated by the competitive pressures discussed earlier. The developing world as a whole spent in the range of US$52 billion on grain imports in 2007, a 10 percent increase from 2006, on the heels of a 36 percent increase from 2005 (Halweil 2007). At the same time, food aid has suffered as relatively fixed budgets encountered rising costs of food.

The supply-side stress on global food prices threatens to intensify far beyond the pressure associated with the scarcity-induced rising cost of oil and natural gas. Though global food supplies have continued to rise in aggregate terms, both grain production and arable land are declining on a per capita basis as the human population approaches 9 billion by mid-century (Halweil 2007). Mounting pressure on land and water supplies is magnified further by climate change (Solomon et al. 2007).

Since the early 1990s, the IPCC has drawn attention to the uneven responsibility and vulnerability associated with anthropogenic climatic change. The vast bulk of greenhouse gas emissions have come from the industrial activity of the world's wealthiest nations (more than 80 percent since the Industrial Revolution), especially the carbon released by the combustion of fossil energy. At the same time, many of the world's poorest nations, and the poorest within them, are projected to be the most immediately and adversely affected.

The Fourth Assessment Report of the IPCC (Solomon et al. 2007) expressed the state of climate science in its strongest terms yet, describing the "warming of the climate system" as being "unequivocal." It also stated that a certain level of warming over the next three decades (roughly 0.2°C per decade) is unavoidable as a result of ocean thermal lag, and that there is therefore a need for climate change adaptation. This is especially so since the climatic changes that we are already committed to loom heavily and unequally, with increasingly varied and extreme conditions expected to impact societies in savannah and tropical lowland regions the most severely. For instance, the IPCC describes Africa as being "one of the most vulnerable continents to climate variability," with 75 to 250 million people expected to face increasing water stress within the next decade, portending huge declines in agricultural yields.

Adaptation to already committed changes cannot be confused with the urgent need to mitigate against more extreme impacts through swift and dramatic reductions in greenhouse gas emissions. In a highly optimistic scenario of aggressive mitigation, the climate change that the world is already committed to would not destabilize temperate agriculture too severely (an area of considerable scientific debate), and the rate of change would be slow enough to plan for a transition to post-fossil energy agriculture. But even in such a scenario, the range of committed change will pose very serious agricultural problems for many of the world's poorest regions and is likely to carry the prospect of deepening food-import dependence on temperate regions—a terribly cruel and regressive outcome, magnified in an era of higher food prices.

Failing major immediate mitigation, the IPCC (Solomon et al. 2007) warns that world will face "abrupt and irreversible changes" to the relative climate stability of the Holocene, with massive declines in agricultural production across much of the developing world and declining aggregate production on a world scale.

Conclusion: Opportunity in Crisis

Given the limits of fossil energy supplies and the fact that industrial agriculture is a primary source of greenhouse gas emissions, it is myopic to assume that industrial production can provide a massive supply side response to the global food crisis. Yet in the near term, strong market signals continue to guide agro-TNCs and industrial agriculture down this unsustainable course. This includes most notably the steady "meatification" of diets and the rising demand and subsidies for biofuels as industrial agriculture gets cast as a partial but significant fix in the desperate search for liquid fuel. In other words, the destructive logic of industrial agriculture is being entrenched amid deepening contradictions while the diffuse and uneven fallout from climate change looms ever larger without serious mitigation efforts.

In the absence of strong anti-systemic movements (i.e., ones capable of presenting a productive alternative), regressive outcomes are likely, and the social fallout associated with rising food prices represents the early stages of this. But there are also potentially highly decentralizing tendencies as the costs of fossil energy and derivatives rise and scarcity sets in, and this is a key part of the transformative opportunity contained in the current system crisis.

As has been emphasized, fossil energy and derivatives have a crucial role in overriding the constraints posed by plant physiology and soil biochemistry, and in substituting technology for human labour, skill, and localized knowledge. Because of this, fossil inputs are implicated in the shifting of value and control away from farmers and into the realms of agro-inputs, traders, processors, distributors, and retailers. In the absence of the biophysical overrides that have sustained industrial agriculture, farming systems will have to become vastly more labour- and knowledge-intensive and locally oriented, and diets vastly less meat-intensive.

Instead of substituting non-renewable fertilizers for the mining of crucial soil nutrients, much more laborious efforts are needed to maintain and build soil fertility, such as intercropping (particularly legumes, which fix atmospheric nitrogen in the soils that gets released to other plants) and managing and digging in agricultural "wastes" such as grain, weed, and especially legume residues. Biological controls for pests, weeds, and disease cycles require more careful and diversified planting, seed saving and selection, and vastly more time-consuming care, such as hand-weeding and inspecting for pests.

There is potential that this lower-input organizing imperative—the re-centring of skill and labour in place of fossil energy, machines, inorganic fertilizers, and chemicals—could generate powerful new openings to return an increasing share of the value within agriculture to farmers and farming communities. Such a hope does not hearken back to some romanticized golden age of the past, and certainly the need to re-substitute labour for machines and fossil inputs will not tend toward greater equity in any inevitable sort of way. Low-input agriculture through millennia was, after all, rarely a farmer-, worker-, or gender-equity paradise, nor was it always sustainable. It is not hard to imagine the ratcheting up of labour exploitation. Indeed, the rise of "Big Organic" in places like California gives some indication of how a labour-intensive, lower-input, and non-toxic but still highly uneven system could emerge if large TNCs are the ones driving the system "fixes," with machines re-substituted with mostly low-paid, non-unionized, and insecure jobs.

Yet where anti-systemic seams created by deepening contradictions are widened from below by progressive farmer and consumer movements, there is hope that more democratic, socially just, and ecologically rational agricultural systems can be built. Again, this does not point to a simple veneration of historic social relations, traditions, and local knowledge. But it does mean re-conceptualizing "modern" agriculture, from approaches based on scale and fossil energy to those based on diversity, complementarity, and respect for ecosystem cycles. This will require major public investment in research, training, extension, and income support, especially in a transitional period and in a changing climate.

Crises can present powerful opportunities for change, and whether the upwelling from below is strong and converging or weak and disparate, a great challenge for agricultural and social scientists is to nourish it and to help foster an understanding of systemic problems and alternatives. The urgency and magnitude is impossible to overstate; as Lester Brown notes, "There's not nearly enough discussion about how people will be fed 20 years from now" (quoted in Leahy 2006).

Notes

1 Although only about 10 percent of all agricultural production in the world is traded across borders, as trade is progressively liberalized world market prices have an increasing impact on prices in domestic markets. International agro-trade is dominated by production from a small number of powerful exporting nations, including the United States and the member states of the EU, where the large majority of global agricultural subsidies are concentrated (Rosset 2006).

2 Geological surveys suggest that mined phosphorous supplies are diminishing quickly. Extracting phosphorus can also pose serious environmental problems, including heavy metal contamination.

3 Natural gas, upon which synthetic fertilizer hinges, also approaches its "peak" in a few decades, though rising consumption levels could push this forward.

4 This is because of the research and development needed in production and processing, the magnitude of land conversion this would entail, and the scale of engineering and technological retrofits that would be required.

Works Cited

Black, Richard (2008). "Shun Meat, Says UN Climate Chief." *BBC News*. 8 September.

BP Global (2008). "Statistical Review of World Energy 2008." London. http://www
.bp.com/liveassets/bp_internet/globalbp/globalbp_uk_english/reports_and_publications
/statistical_energy_review_2008/STAGING/local_assets/downloads/pdf/
statistical_review_of_world_energy_full_review_2008.pdf.

Brown, Lester (1996). *Tough Choices: Facing the Challenge of Food Scarcity*. New
York: W.W. Norton.

———— (2007). "Distillery Demand for Grain to Fuel Cars Vastly Understated: World
May Be Facing Highest Grain Prices in History." Washington, DC: Earth Policy
Institute. http://www.earth-policy.org/Updates/2007/Update63.htm.

The Economist (2007). "Cheap No More." *The Economist*. 8 December: 81–83.

———— (2008). "Oil: Well Prepared." *The Economist*. 8 November. http://www
.economist.com/finance/displaystory.cfm?story_id=12564013.

Ferrett, Grant (2007): "Biofuels 'Crime against Humanity.'" *BBC News*. 27 October.

Food and Agriculture Organization of the United Nations (2002). "World Agriculture:
Towards 2015/2030—Summary Report." Rome. ftp://ftp.fao.org/docrep/
fao/004/y3557e/y3557e.pdf.

———— (2008). "The State of Food Insecurity in the World, 2008." Rome. ftp://ftp
.fao.org/docrep/fao/011/i0291e/i0291e00.pdf.

Foster, John B., Hannah Holleman, and Robert McChesney (2008). "The US Imperial
Triangle and Military Spending." *Monthly Review* 60, no. 5: 1–19.

Goodland, Robert, and David Pimentel (2000). "Environmental Sustainability and
Integrity in the Agriculture Sector." In *Ecological Integrity: Integrating Environment,
Conservation, and Health*, ed. David Pimentel, Laura Westra, and Reed F. Noss.
Washington, DC: Island Press.

Halweil, Brian (2007). "Grain Harvest Sets Record, but Supplies Still Tight." Washington, DC: WorldWatch Institute.

Hamilton, Tyler (2008). "How Low Can Oil Go?" *Toronto Star*. 15 November.

Heinberg, Richard (2005). *The Party's Over: Oil, War, and the Fate of Industrial Societies,* 2nd ed. Gabriola Island, BC: New Society Publishers.

International Energy Agency (2007). "Key World Energy Statistics." Paris.
http://www.iea.org/textbase/nppdf/free/2007/Key_Stats_2007.pdf.

Leahy, Stephen (2006). "Population: Global Food Supply Near the Breaking Point."
Inter-Press Service News Agency. 17 May.

Monbiot, George. (2007). "The Western Appetite for Biofuels Is Causing Starvation in
the Poor World." *The Guardian*. 6 November. http://www.guardian.co.uk/
commentisfree/2007/nov/06/comment.biofuels.

Nierenberg, Danielle (2005). "Happier Meals: Rethinking the Global Meat Industry."
WorldWatch Paper no. 171. Washington, DC: WorldWatch Institute.

Oil & Gas Journal (2006). "Worldwide Look at Reserves and Production." *Oil & Gas
Journal* 104, no. 47: 24–25.

Patzek, Tad W., and David Pimentel (2006). "Thermodynamics of Energy Production
from Biomass." *Critical Reviews in Plant Sciences* 24, no. 5–6: 329–64.

Pimentel, David, and Tad W. Patzek (2005). "Ethanol Production Using Corn, Switch-
grass, and Wood; Biodiesel Production Using Soybean and Sunflower." *Natural
Resources Research* 14, no. 1: 65–76.

Righelato, Renton, and Dominick V. Spracklen (2007). "Carbon Mitigation by Biofuels or by Saving and Restoring Forests?" *Science* 319: 902.

Rosset, Peter (2006). *Food Is Different: Why We Must Get the WTO Out of Agriculture.* London: Zed Books.

Sample, Ian (2007). "Global Food Crisis Looms as Climate Change and Population Growth Strip Fertile Land." *The Guardian*. 31 August.

Solomon, S., D. Qin, M. Manning, Z. Chen, M. Marquis, K.B. Averyt, M. Tignor, and H.L. Miller, eds. (2007). *Climate Change 2007: The Physical Science Basis, Contribution of Working Group I to the Fourth Assessment Report of the Intergovernmental Panel on Climate Change.* Cambridge: Cambridge University Press.

Weis, Tony (2007). *The Global Food Economy: The Battle for the Future of Farming.* London: Zed Books.

WorldWatch Institute and Center for American Progress (2006). "American Energy: The Renewable Path to Energy Security." Washington, DC.

CHAPTER 12

Setting the Global Dinner Table
Exploring the Limits of the Marketization
of Food Security

Noah Zerbe

Fifty years ago, the international community faced a food crisis. Growing world population, particularly in South Asia, threatened to outstrip the world's food supply, presenting the danger of realizing Malthus's dark future. National and international agricultural research was mobilized in what came to be known as the "Green Revolution," a concerted international plant breeding program intended to develop new plant varieties that were hardier, took less time to grow, and were more responsive to chemical inputs. Although the Green Revolution had important—though unintended—environmental, economic, and social consequences, it also improved crop yields in the areas it was deployed, pushing back the specter of Malthusian famine.

The 2008 global food crisis—the global food situation that has ensued clearly warrants use of the perhaps overused term—presents a similar, grave threat, and some have argued that agricultural biotechnology can again provide relief. While food prices have steadily increased since 2000, a sharp spike in global food prices can be seen between 2006 and 2008 (Steinberg 2008). Although by early 2009 international food prices retreated from their peak levels of a year earlier (largely due to the global economic crisis that trimmed global demand for primary commodities across the board) global food prices remained well above average pre-crisis levels.

In the context of the global food crisis, food security has re-emerged as a central issue in the global political economy. Dramatic increases in the prices of basic foodstuffs have sparked protests and food riots in more than fifty countries around the world. The Food and Agriculture Organization of the United Nations

161

(FAO) projected that thirty-six countries would require food aid in the winter of 2008–09 (FAO 2008). The World Food Programme (WFP) estimated that it would require an additional US$755 million to feed the estimated 73 million people across seventy-eight countries relying on food aid in 2008. The increased cost of food aid, according to the WFP, was driven both by higher food prices and by higher transportation costs associated with the spike in global oil prices (UN News Center 2008). While the causes of the crisis are diverse, the crisis itself has clearly forced the world's poor into an increasingly stark reality.

In this context, proponents of agricultural biotechnology have argued that the technology presents a simple and straightforward solution to the crisis—one that might solve the supply-side issues often considered to be at the heart of such crises. But advocating agricultural biotechnology as the solution to the current crisis presupposes the nature of the crisis itself. In this chapter, I explore the causes of the global food crisis, arguing that the current crisis should not be seen merely as the result of growing demand or insufficient production. While this is certainly a part of the story, as is often the case, the reality is much more complex. After outlining the causes of the global food crisis, I consider proposed solutions. I contend that while agricultural biotechnology may be able to increase production, the complexity of the crisis requires a more nuanced solution. In particular, I argue that because the origins of the crisis rest primarily in the marketization of food security rather than in the failure of global production, solutions to the crisis must necessarily include efforts to re-embed food security in social relations.

Causes of the Global Food Crisis

Speaking in May 2008, then US President George W. Bush outlined his understanding of the global food crisis. According to Bush, "There are 350 million people in India who are classified as middle class. That's bigger than America. Their middle class is larger than our entire population.[1] And when you start getting wealthy, you start demanding better nutrition and better food so demand is high and that causes the price to go up" (cited in Baruah 2008). In his testimony before the US Senate Foreign Relations Committee, Peter McPherson offered a more comprehensive assessment of the causes of the current global food crisis. According to McPherson, "a number of factors have contributed to the great jump in food prices, but the problem has been long in the making" (McPherson 2008). Those long-term causes, in McPherson's estimation, include:

- cuts to agricultural investment by national governments and international organizations and related cuts in foreign agricultural assistance programs over the past thirty years;

- increasing incomes, particularly in Asia, leading to changes in diet; and
- state interference in the free market, particularly through agricultural subsidies.

A number of shorter-term causes are also important. These include:

- increasing demand for biofuels, leading to higher prices for primary food commodities, particularly corn;
- higher fuel and energy prices increasing the cost of agricultural inputs (fertilizers), farm operation, and food transport;
- instability introduced by the breakdown of global credit markets;
- declines in global grain reserves from 100 days in 2000 to 55 days by May of 2008, leading to greater sense of risk and market instability;
- the Australian drought; and
- food export restrictions imposed by many countries in an attempt to cope with increasing food prices (ibid., 1–2).

McPherson's analysis of the causes of the food crisis seems comprehensive, highlighting the complexity of the current situation, which appears to represent the "perfect storm" of underproduction and excessive regulation.

Although this perfect storm scenario makes sense on the surface, I argue that framing the problem as one of insufficient production and overregulation leads to problematic solutions. In terms of regulation, the agricultural sector should be characterized neither as over- nor under-regulated, but as both. In the global North, it may be accurate to argue that excessive state intervention has altered food markets in profound ways. The massive agricultural subsidies afforded farmers in Japan, Europe, and the United States have lowered global food prices, encouraging overproduction of certain grain crops, which has resulted in dumping and the destruction of agricultural production and development prospects in the global South (Thompson and Stoneman 2007). But Northern subsidies have been largely excluded from international trade agreements despite the fact that such subsidies are valued at more than $300 billion per year—six times the total value of development aid (Clapp 2006). The impact of these subsidies was noted by former US President Bill Clinton, who, just before leaving office in 2001, noted that, "If the wealthiest countries ended our agricultural subsidies, leveling the playing field for the world's farmers, that alone would increase the income of developing countries by US$20 billion a year" (cited in Thompson 2004).

But in the global South, agricultural production has been extensively deregulated and subject to the discipline of the market since the 1980s. The uneven process of deregulation and liberalization, under which the poor in the global

South have been subject to market forces while the relatively wealthy in the global North have been protected against market discipline, has rightly been critiqued. But the question that arises—and to which I return below—is how to move forward in addressing this inequality. Should global agriculture be liberalized or re-regulated? And more importantly, what should be the goals of public policy and governance in agriculture? How best, in short, can we achieve food security?

First, an important caveat: Increasing the productivity of rural smallholders in the global South must surely be an important part of any food security strategy. The rural poor represent upward of three-quarters of the world's most impoverished people, defined by the World Bank as those living on less than US$1 per person per day. The vast majority are either small producers or those who derive their livelihoods from agricultural production. The fundamental question we must address, therefore, is how to improve their livelihoods. What changes are necessary to improve the plight of the world's rural poor? How can their income and quality of life be bettered?

In this context, the global food crisis would seem to present both challenges and opportunities. Indeed, a report published by the Organization for Economic Co-operation and Development concludes, "The curse of higher food prices can be turned into a blessing if African agriculture finally becomes a business" (Wolter 2008). Higher agricultural prices would appear to be a boon for rural smallholders, a group traditionally at best overlooked and at worst penalized by development strategies. Smallholder farmers who produce crops for sale may indeed be able to benefit from higher prices for the food they produce. There is, however, reason to suspect this may not turn out to be the case.

The Limits of Marketization of Food Security: Lessons from the Zimbabwean Seed Industry

Framing the global food crisis as the result of underproduction and market interference necessarily leads to two fundamental conclusions. First, agricultural production needs to be expanded, preferably quickly, in order to have enough food to feed the world. Second, food markets need to be deregulated in order to permit market forces to encourage the expansion of agricultural production.

This framing of the food crisis represents a fundamental break from the historical understanding of the nature of food production. Prior to the rise of neoliberalism in the 1980s, food production was dominated by the principle of national self-sufficiency. Food security was conceptualized primarily at the national level, with some recognition of regional food security such as at the level of the Southern African Development Community. Most food was produced and consumed locally. Local food programs and national grain reserves were often instituted with funding from the international development agencies. Zimbabwe's post-independence plant-breeding effort—one of the most advanced plant-breeding initiatives in Africa and, by nearly all accounts, a dramatic success—was initially financed, in part, by the World Bank.

But under the logic of neo-liberalism, the free market would guarantee food security through comparative advantage—lowering prices of consumers while raising incentives for producers. In Southern Africa, comparative advantage meant that some countries, like Botswana, would produce little or no food but would instead purchase the cheap grain on international markets with revenues earned from other exports—in Botswana's case, diamonds. Countries like Zimbabwe, with comparative advantage in grain production, would become major food exporters—the grain basket for the region—by removing inefficient and corrupt government intervention from agricultural markets and allowing markets to encourage efficient production through price signals. The market, rather than the state, thus becomes the primary guarantor of food security, a phenomenon we might rightly conceptualize as the marketization of food security.

In the case of Zimbabwe, structural adjustment was particularly successful in reducing government interference in the market. Between 1991 and 1995, the government commercialized agricultural marketing boards, liberalized agricultural finance, and cut spending on agricultural development. It also privatized its historical system of seed distribution so that the market would assume responsibility for the development of new seed varieties. But in Zimbabwe, the effect of marketizing food security was devastating. As commercial marketing boards were privatized in the early 1990s, producer prices collapsed. For maize, the staple food crop, local producer prices often failed to cover the cost of production. Private traders who ventured into remote areas could offer prices well below the historical "floor prices" offered by the national Grain Marketing Board (GMB).[2] The elimination of pan-territorial pricing subjected all farmers to the discipline of the market. Consequently, it also exacerbated regional inequalities, as those living closer to Harare, Zimbabwe's capital, could demand higher prices on wholesale markets because transportation prices were lower (Jones 1992). As a result of the elimination of pan-territorial pricing, all farmers, regardless of their proximity to the most profitable urban markets, were made increasingly vulnerable to dramatic shifts and instabilities in global commodity prices.

In Zimbabwe, the marketization of food security simultaneously introduced greater instability in producer prices. The floor prices established under the old system of the GMB were set early in the season, allowing farmers to plan production based on steady and predictable future market prices. The elimination of the GMB meant that farmers were subject to the vagaries of the market, unable to depend on or plan for specific prices, and increasingly subject to fluctuations based on weather, international markets, or speculation.[3]

A report by the United Nations Conference on Trade and Development confirms the relevance of the Zimbabwean experience for other African economies. According to the report, structural adjustment programs saw "growing profit margins for private traders at the expense of farmers." The dismantling of inefficient marketing boards only occasionally led to more competitive or efficient markets (e.g., cotton). The study concluded, "Africa's experience with agricultural reform is yet another example of a 'big bang' liberalization without preparing for institutions and the infrastructure needed for markets to perform effectively" (UNCTAD 1997, 7). The net result is a combination of state *and* market failure.

The impact of structural-adjustment-inspired privatization on Zimbabwe's seed industry is particularly telling. Along with South Africa, which has much higher levels of overall development and a much larger population, Zimbabwe historically boasted the most successful seed industry in the region. The formal seed sector in Zimbabwe, composed of more than twenty companies, produced more than 70,000 metric tons of improved seed for the domestic market annually before the collapse of Zimbabwe's agricultural sector in the early 2000s.[4] In maize seed in particular, Zimbabwe enjoyed particular success, with near universal adoption of hybrid maize cultivars developed by the state for local growing conditions.

The success of Zimbabwe's seed system was largely the result of the unique public–private partnership developed in the country between the 1940s and 1991. Early in Zimbabwe's history, the colonial government of Southern Rhodesia, owing in large part to the nature of settler colonialism and the political importance of white settler farmers, put a great deal of effort into the development of improved cultivars for commercial agricultural production. In the 1930s, government breeders began to experiment with hybrid maize, some varieties of which performed well even under adverse farming conditions. Lacking the capability to produce and distribute hybrid maize seed on a commercial level, the government entered into an agreement with the Seed Maize Association (SMA) of Zimbabwe.[5] Under the terms of the agreement, the government would provide improved maize varieties to the SMA, which would produce and market the improved seed at set prices negotiated with the government.

The original agreement between the government and the SMA was a great success, and Southern Rhodesia quickly became a major seed producer and exporter. In 1949, Southern Rhodesia became the second country (after the United States) to produce double cross hybrid seed, designated SR1, from locally developed inbred lines. The position of Southern Rhodesia as the lead-ing regional seed provider was solidified throughout the 1950s, with the con-tinual release of new hybrid cultivars culminating in the release of SR52 in 1960 and the R200 series thereafter (Rusike 1995).

In an attempt to attain self-sufficiency in maize production, the post-inde-pendence government instituted a series of policies designed to promote the use of hybrid maize among smallholder farmers. It expanded joint research efforts between the government and the SMA (which became Seed Coop in 1979). The primary focus of the post-independence research effort was on the development of cultivars designed for low rainfall and marginal farming areas. Extension services touted the benefits of new hybrid maize varieties to small-holders, while the government simultaneously prohibited the sale of open-pol-linated varieties.[6] It also maintained price controls on maize seed instituted by the colonial government. The combination of price controls and direct distri-bution from Seed Coop to farmers, bypassing middlemen and seed traders com-mon in other types of seed distribution networks, meant that Zimbabwe enjoyed some of the lowest seed prices in the world. The results of its efforts were dra-matic: within ten years of the post-independence government coming to power, almost all maize grown in Zimbabwe was produced from certified, hybrid seed (Muir and Blackie 1994; Cromwell 1996).[7]

The introduction of structural adjustment, however, undermined govern-ment and Seed Coop success in the production and provision of improved vari-eties to smallholder farmers. As part of conditions placed in the loan agreement between the Zimbabwe government and the World Bank, the government was to sever its relationship with Seed Coop and eliminate price controls on seed. It was hoped that such policies would open the way for greater competition in the seed and agricultural inputs sectors (Rusike 1995).[8]

After the government was forced to terminate its relationship with Seed Coop, the cooperative transformed itself into a private seed company known as Seed Co. As a private seed breeder, Seed Co maintained its own research sta-tions and proprietary lines (Seed Co 2001, 3). Although as Seed Coop it had maintained independent research stations since 1973,[9] Seed Co now had primary responsibility for developing new cultivars. Seed Co subsequently expanded its operations and up to the late 1990s maintained a key role in regional seed mar-kets, particularly in Zambia, Mozambique, Botswana, Malawi, and South Africa.[10] Other companies, including Monsanto, Pioneer, and Pannar, entered the Zimbabwean market, but failed to fundamentally challenge the position of Seed Co.

The introduction of competitive markets in seed and inputs was supposed to increase seed supplies and decrease production costs. Any cost increases would, according to the World Bank, be offset by increases in commodity prices received by farmers. However, pre-adjustment predictions typically underestimated inflation in the general economy,[11] which eroded the real purchasing power of smallholder farmers. In the first five years of reform, fertilizer prices reflected the 400 percent currency devaluation—quickly pricing the smallholder out of the market because fertilizer chemicals are imported. According to Sibanda, the high cost of inputs—especially fertilizer—became a major constraint for over 50 percent of smallholder farmers (Sibanda 2001). In 1997–98, for example, the price of fertilizer increased by a factor of seven in one year; chemicals for pesticides increased over 100 percent, and the cost of electricity doubled. Prices for insecticides rose as much as 118 percent, and seed dressing 107 percent in the same year (Commercial Farmers Union of Zimbabwe 1998). The increasing cost of inputs forced many smallholder farmers to cut investment, leading to an overall decline in total maize yields of 2.75 percent despite an increase of 2.52 percent in total area farmed to maize in the 1990s. This decline in yields stood in sharp contrast to the country's first decade of independence, when extensive state investment in smallholder agricultural production led to increases in maize yields of 2.21 percent despite a small decline in the total area dedicated to maize production (Smale and Jayne 2003, 49).

The tragic irony of the privatization and liberalization of Zimbabwe's seed industry is the fact that policies intended to promote agricultural production as a mechanism for addressing rural poverty actually wound up exacerbating rural poverty, undermining smallholder production, and reducing food security. Zimbabwe's pre-adjustment seed program illustrated the potential benefits of pro-poor agricultural research accompanied by effective distribution channels. It also demonstrated the important role the state can play in guaranteeing national food security.

The Promise of Agricultural Biotechnology

Despite its diverse causes, the contemporary global food crisis has been presented as a development that necessitates a rethinking of the potential benefits of agricultural biotechnology. Indeed, for its proponents, increasing food prices give reason to pause and reconsider consumer resistance to the technology. Biotechnology potentially represents a relatively straightforward technical fix to the problem of higher food prices. Technological innovation may, for example, allow farmers in the developing world to grow higher-yielding crops under less favourable conditions in shorter growing seasons. The combined impact, it is hoped, would be a dramatic increase in total yields, paralleling similar advances that took place in the 1960s during the Green Revolution.

The success of Zimbabwe's seed-breeding program suggests that the development of new plant varieties may indeed represent a boon to agriculture. Certainly the Green Revolution demonstrates the technical possibilities. According to the International Food Policy Research Institute, rice and wheat yields doubled as a result of the Green Revolution (IFPRI 2002, 2). However, as with all technical and economic changes, the benefits were not evenly distributed across society. Indeed, in some instances, the Green Revolution increased social, political, and economic inequalities and resulted in profound negative externalities in the environment. As Middendorf et al. caution, "The Green Revolution provides an important lesson with respect to the application of new agricultural biotechnologies: the ideology of inevitable technological progress excludes consideration of the distributional and environmental consequences of such efforts" (Middendorf et al. 1998, 93–94).

A study by the FAO similarly concludes that the higher yields resulting from the Green Revolution were accompanied by a number of unintended consequences, including a concentration of land ownership, a shift in the gendered dynamic of agricultural production against women, increasing inequality between early and late adopters of seeds and other inputs, and loss of biodiversity (FAO 2001, 7).

The unintended consequences arising from the Green Revolution are likely to be exacerbated by the Gene Revolution. Unlike the Green Revolution technologies, which were developed by nonprofit researchers and distributed largely through governmental networks, the current Gene Revolution is largely a privately funded, for-profit endeavour. This may be an appropriate model in the global North, where commercial, capital-intensive agriculture has been the norm for quite some time. But for the vast majority of smallholder farmers in the global South, these capital-intensive farming methods make little sense. Adoption of capital-intensive farming technologies in the global South would likely result in the replication of patterns of socio-economic inequality that followed the Green Revolution. This is a real concern, as even the Syngenta Foundation concedes:

> In a socially and politically defective setting [biotechnology] can hardly bring about improvements in the condition of the poor. Where land ownership and tenancy systems, access to extension services, credit, marketing channels, as well as new technologies are governed by a socio-political power structure that favours only a small minority technological progress cannot possibly be neutral in its impact....In short, the developmental impact of recombinant genetics and biotechnology is only as good as the socio-political soil in which they are planted. Any technical advance, progress in genetics included, can only benefit those who understand the technology and are able to apply it. Every

> restriction on access, be it lack of schooling, or feudal power structures, can
> have the effect of aggravating income disparities—pronouncedly so when the
> technology is very potent. Unless social reforms are introduced and reinforced
> with supportive measures that also enable the middle and lower strata of soci-
> ety to gain their share step by step, technological innovations actually work
> against the goal of breaking down inequalities. (Syngenta Foundation 2002)

Syngenta correctly recognizes the possibility of increasing inequality as a result
of the adoption of agricultural biotechnology. But while Syngenta is correct to
acknowledge the problems of "restrictions on access," it incorrectly places the
primary responsibility for the problem on the predatory state. More accurately,
the most important restriction on access to any technology is not the predatory rule of the state but the natural functioning of the market. The central
objective of any market is the allocation of scarce resources (land, credit, technology, inputs, and so on) to particular groups based on an arbitrary measure
of worthiness (money).

Further complicating the question of adoption is the direction of research.
Much has been made of the potential of biotechnology to result in the creation
of new crop varieties more suited for local growing conditions in the developing world. Particular emphasis has been placed on crops engineered for higher
yields, greater stress tolerance (cold tolerance, heat tolerance, drought tolerance,
saline tolerance), improved nutritional content (Golden Rice, high-protein cassava), and other benefits (malaria-vaccine bananas and cassava). Such crops
may indeed bring real benefits. But the current generation of commercially available biotechnology is concentrated in just four crops (corn, soy, cotton, and
canola) and two traits (herbicide tolerance and drought resistance) (James 2007).

The concentration of research in particular crops and traits is largely a reflec-
tion of the commercial focus of the research. The current generation of research
in agricultural biotechnology is directed by the private sector, and there is little
reason to believe that the private sector has significant interest in developing
crops specifically targeted for smallholder farmers in the global South. We are
far more likely to see meat genetically engineered to be lower in calories or cho-
lesterol than we are to see drought-tolerant cassava. This should not be a surprise;
private sector research is market-driven and highly responsive to consumer
demand. Demands unable to be expressed through the market (such as those of
smallholder farmers with little money to purchase new seed every year) are
unlikely to receive any significant attention from private-sector research. In this
sense, reliance on the market for the development of new crop varieties for poor
farmers mirrors the failure of pharmaceutical companies to direct research into
so-called "neglected diseases." As Trouiller et al. observe, "Today's R&D-based
pharmaceutical industry is reluctant to invest in the development of drugs to

treat the major diseases of the poor, because return on investment cannot be guaranteed. With national and international politics supporting a free market-based world order, financial opportunities rather than global health needs guide the direction of new drug development" (Trouiller et al. 2001, 945).

Just as in the case of limited pharmaceutical research into neglected tropical diseases (Burri 2004; Manderson et al. 2009), the limited research in crop varieties of particular utility to smallholder farmers is not primarily a function of the technology itself but rather of the commercial focus of profit-driven research. The state thus has a role to play. But in Zimbabwe, the historical climate associated with neo-liberalism eviscerated the state. Government budgets were slashed, and plant breeding increasingly fell under the nearly exclusive purview of the market. Despite important advances made by plant breeders, national seed systems across the global South were privatized in the 1980s and 1990s under the belief that the market would be more responsive to farmer demand. And it was responsive to farmer demand but not to the demand of all farmers equally.

At its best, technological innovation has the potential to address the real problems faced by smallholder farmers, assuming that this research is directed in a pro-poor direction. But private research is rarely targeted at the poor, especially those in rural areas, the population most at risk. Research on traits demanded by smallholder farmers must proceed, but cautiously, cognizant of unintended consequences associated with technological change. At its worst, the emphasis on technological solutions to the global food crisis represents the type of anti-politics decried by Ferguson's analysis of development in Lesotho; a depoliticization and dehistoricization of the current crisis (Ferguson 1990).

Conclusion: Governance and Decommodification

Writing of the 1940s, Karl Polanyi noted the dangers associated with the "commodity fiction"—the treatment of land, labour, and money as commodities produced for sale on the market. Polanyi lamented the destructive impulse of such a system, noting,

> To allow the market mechanism to be sole director of the fate of human beings and their natural environment, indeed, even of the amount and use of purchasing power, would result in the demolition of society. For the alleged commodity "labor power" cannot be shoved about, used indiscriminately, or even left unused, without affecting also the human individual who happens to be the bearer of this peculiar commodity.... Robbed of the protective covering of cultural institutions, human beings would perish from the effects of social exposure; they would die as the victims of acute social dislocation through vice, perversion, crime, and starvation. Nature would be reduced to its elements,

neighborhoods and landscapes defiled, rivers polluted, military safety jeopardized, the power to produce food and raw materials destroyed. Finally, the market administration of purchasing power would periodically liquidate business enterprise, for shortages and surfeits of money would prove as disastrous to business as floods and droughts in primitive society.... But no society could stand the effects of such a system of crude fictions even for the shortest stretch of time unless its human and natural substance as well as its business organization was protected against the ravages of this satanic mill. (Polanyi 1957, 73)

Polanyi's purpose was to analyze the constant struggle between market regulation and state regulation—a struggle played out though the double movement. Ultimately, the solution to the problem of the commodity fiction centred on recognizing the unique nature of fictitious commodities.

While Polanyi never dealt explicitly with food as a commodity, a compelling argument could be made regarding the unique nature of food. Indeed, over the past thirty years, the guarantees for food security have increasingly become the purview of the market. Structural adjustment and privatization resulted in the transfer of responsibility for food security in general (and seed production in particular) from the state to the market—a process I term the marketization of food security. Even where the state was able to provide high-quality, low-cost seed to small farmers, as was the case in Zimbabwe, the drive for small government and greater efficiency resulted in the dismantling of seed systems and an introduction of the market logic of production into seed networks. The result, as noted above, was the development of a two-tiered seed system under which farmers who could express demand through the market continued to receive high-quality seed developed using novel techniques in molecular agricultural biotechnology and distributed through commercial channels, while the needs of smallholder farmers who could not afford to purchase commercial seed were essentially ignored.

The alternative to the commodity fiction and the marketization of food security may rest in the moral economy. As articulated by economist E.P. Thompson, the "old moral economy of provision" emphasized the common well-being of society, which placed limits on the operation of the market (Thompson 1971). The moral economy model outlined by Thompson thus presents an important counter to the logic of the market.

Perhaps food is unique among commodities in that it is fundamentally necessary for human existence. In this respect, any solution to the current food crisis should begin with the premise that food should not be subject merely to regulation of the market; that food security should be a central goal of state policy. Technological innovation has a role to play in expanding agricultural production. The market, too, has an important role. But relying solely on technological innovation, particularly market-based innovation, misunderstands the social context and historical realities of the current crisis. Getting the politics, sociology, and economics right is just as important—perhaps more important—than getting the prices right.

Notes

1 China's growing middle class is often added to the equation, suggesting that the new middle class in China and India account for the bulk of the price increases. Daryll Ray offers a forceful counter-argument to this assertion (Ray 2008).

2 There were, to be certain, problems with grain marketing boards in Africa. But a study by Lele and Christiansen concludes that the major problems normally associated with marketing boards were the result of their undercapitalization and over-extension (Lele and Christiansen 1989). At a minimum, it should be recognized that some agricultural sub-sectors, particularly those high-volume, low-priced commodities grown in marginal and remote regions, are often unattractive to private traders (Gibbon, Havnevik, and Hermele 1993).

3 It should be noted that Northern farmers still benefit from these types of floor prices intended to mitigate against market fluctuations and uncertainties associated with agricultural production.

4 Vincent Gwarazimbaincent, General Manager of Zimbabwe Seed Trade Association. Interviewed by the author, Harare, 3 April 2001.

5 The Seed Maize Association was originally founded in 1940 with the purpose of producing seed for its members. The SMA centralized seed multiplication, which historically had been done by individual members, thus providing economies of scale in seed production and freeing its members to focus on production of maize for the market. In 1979, it was renamed Seed Coop, which became Seed Co after the introduction of structural adjustment in 1991.

6 In the face of the collapse of agricultural production in the early 2000s, the Government of Zimbabwe rescinded its longstanding prohibition on the sale of open-pollinated varieties of maize, which had remained under cultivation in some parts of the country despite the official ban.

7 Anthony Mashiringwani, Seed Breeder, Department of Research and Specialist Services, Ministry of Lands and Agriculture, Government of Zimbabwe. Interviewed by the author, Harare, 27 April 2001.

8 Ibid.; Barry McCarter, General Manager, Seed Co. Interviewed by the author, Harare, 24 May 2001.

9 In an effort to expand the number of seed-line varieties under research and cultivation, Seed Coop established the Rattray Arnold Research Station outside Harare in 1973. Funded by Seed Coop, the station operated in cooperation with (but independently from) government research and breeding programs. The station was intended to complement government research efforts.

10 In the 1990s, Seed Co developed production facilities in Botswana, Malawi, Mozambique, and Zambia, and operated facilities in South Africa through a joint agreement with Syngenta. It also regularly exported seeds to Swaziland, Tanzania, Kenya, Cameroon, Angola, and the DR Congo (McCarter interview 2001; Seed Co 2001).

11 Prior to the introduction of structural adjustment, inflation averaged approximately 13.4 percent per year (between 1980 and 1990). Following adjustment, inflation increased to 26.7 percent between 1991 and 1995, and soared to 40 percent between 1995 and 1997.

Works Cited

Baruah, Amit (2008). "Worry for Bush: Indians Spend $20m on Pet Food." *Hindustan Times*. 6 May.

Burri, Christian (2004). "High Time to Take Action: Research on Neglected Diseases." *Bulletin von Medicus Mundi Schweiz* 92 (April). http://www.medicusmundi.ch/mms/services/bulletin/bulletin200401/kap01/08burri.html.

Clapp, Jennifer (2006). "WTO Agriculture Negotiations: Implications for the Global South." *Third World Quarterly* 27, no. 4: 563–77.

Commercial Farmers Union of Zimbabwe (1998). "Survey of Input Costs and Recent Input Cost Increases." Harare, January. Unpublished report.

Cromwell, Elizabeth (1996). "Governments, Farmers, and Seeds in a Changing Africa." London: CAB International and Overseas Development Institute.

Ferguson, James (1990). *The Anti-Politics Machine: "Development," Depoliticization, and Bureaucratic Power in Lesotho*. Cambridge: Cambridge University Press.

Food and Agriculture Organization of the United Nations (2001). "Potential Impacts of Genetic Use Restriction Technologies (GURTs) on Agricultural Biodiversity and Agricultural Production Systems." CGRFA/WG-PGR-1/01/7. http://www.fao.org/waicent/faoinfo/agricult/agp/agps/pgr/itwg/pdf/p1w7e.pdf.

———— (2008). *Crop Prospects and Food Situation*, no. 4 (October). http://www.fao.org/docrep/011/ai473e/ai473e00.htm.

Gibbon, Peter, Kjell Havnevik, and Kenneth Hermele (1993). *A Blighted Harvest: The World Bank and African Agriculture in the 1980s*. Trenton, NJ: Africa World Press.

International Food Policy Research Institute (2002). "Green Revolution: Curse or Blessing?" Washington, DC. http://www.ifpri.org/pubs/ib/ib11.pdf.

James, Clive (2007). "Global Status of Commercialized Biotech/GM Crops: 2007." Ithaca, NY: International Service for the Acquisition of Agri-Biotech Applications.

Jones, Stephen (1992). "Dilemmas of Agricultural Marketing Reform under Structural Adjustment in Zimbabwe: Maize Marketing Policy." Unpublished report.

Lele, Uma, and Robert Christiansen (1989). "Markets, Marketing Boards, and Co-operatives in Africa: Issues in Adjustment Policy." MADIA Discussion Paper no. 11. Washington, DC: World Bank.

Manderson, Lenore, Jens Aagaard-Hansen, Pascale Allotey, Margaret Gyapong, and Johannes Sommerfeld (2009). "Social Research on Neglected Diseases of Poverty: Continuing and Emerging Themes." *Neglected Tropical Diseases* (February). http://www.plosntds.org/article/info:doi%2F10.1371%2Fjournal.pntd.0000332.

McPherson, Peter (2008). "The Global Food Crisis: Causes and Solutions." Statement before the US Senate Foreign Relations Committee. 14 May.

Middendorf, Gerald, Mike Skladany, Elizabeth Ransom, and Lawrence Busch (1998). "New Agricultural Biotechnologies: The Struggle for Democratic Choice." *Monthly Review* 50, no. 3: 85–96.

Muir, Kay, and Malcolm Blackie (1994). "The Commercialization of Agriculture." In *Zimbabwe's Agricultural Revolution*, ed. Mandivamba Rukuni and Carl Eicher. Harare: University of Zimbabwe Press.

Polanyi, Karl (1957). *The Great Transformation: The Political and Economic Origins of Our Time*. Boston: Beacon Press.

Ray, Daryll (2008). "Data Show that China's More Meat-based Diet is NOT the Cause of Ballooned International Corn Prices?" Knoxville: University of Tennessee Institute of Agriculture. http://agpolicy.org/weekcol/408.html.

Rusike, Joseph (1995). "An Institutional Analysis of the Maize Seed Industry in Southern Africa." Unpublished Ph.D. dissertation. East Lansing: Department of Agricultural Economics, Michigan State University.

Seed Co (2001). "Growing into Africa: Seed Co Limited 2001 Annual Report." Harare.

Sibanda, Arnold (2001). "The Role of the State with Respect to Agriculture, Trade Liberalization, Financial Reform, and Labour Markets." Paper presented to the Structural Adjustment Participatory Review Initiative (SAPRI), Poverty Reduction Forum. Harare: Structural Adjustment Participatory Review International Network. 9–10 April.

Smale, Melinda, and Thom Jayne (2003). "Maize in Eastern and Southern Africa: 'Seeds' of Success in Retrospect." Discussion Paper No. 97. Washington, DC: International Food Policy Research Institute. http://www.ifpri.org/divs/eptd/dp/papers/eptdp97.pdf.

Steinberg, Stefan (2008). "Financial Speculators Reap Profits from Global Hunger." Montreal: Centre for Research on Globalization. 24 April. http://global research.ca/index.php?context=va&aid=8794.

Syngenta Foundation (2002). "The Socio-Political Impacts of Biotechnology in Developing Countries." Basel. http://www.syngentafoundation.com.

Thompson, Carol (2004). "Globalizing Land and Food in Zimbabwe: Implications for Southern Africa." *African Studies Quarterly* 7, no. 2–3. http://web.africa.ufl .edu/asq/v7/v7i2a10.htm.

Thompson, Carol, and Colin Stoneman (2007). "Trading Partners or Trading Deals? The EU and USA in Southern Africa." *Review of African Political Economy* 34, no. 112: 227–45.

Thompson, E.P (1971). "The Moral Economy of the English Crowd in the Eighteenth Century." *Past and Present* 5: 75–136.

Trouiller, Patrice, Els Torreele, Piero Olliaro, Nick White, Susan Foster, Dyann Wirth, and Bernard Pécoul (2001). "Drugs for Neglected Diseases: A Failure of the Market and a Public Health Failure?" *Tropical Medicine and International Health* 6, no. 11: 945–51.

UN Conference on Trade and Development (1997). "Trade and Development Report 1997: Globalization, Distribution and Growth". Geneva. http://www.unctad.org/ en/docs/tdr1997_en.pdf.

UN News Center (2008). "UN Food Aid Agency Appeals for $500 Million to Offset Soaring Prices." 24 March. http://www.un.org/apps/news/story.asp?NewsID= 26071&Cr=wfp&Cr1=prices.

Wolter, Denise (2008). "Higher Food Prices—A Blessing in Disguise for Africa?" *OECD Policy Insights* 66 (May). http://www.oecd.org/dataoecd/43/47/40986119.pdf.

PART 4

Strategies to Promote Food Security
and Sustainable Agriculture:
The Way Ahead

CHAPTER 13

A Stronger Global Architecture for Food and Agriculture

Some Lessons from FAO's History and Recent Evaluation

Daniel J. Gustafson and John Markie

The introductory note to the joint Centre for International Governance Innovation–International Food Policy Research Institute workshop on International Governance Responses to the Food Crisis that led to this volume states that:

> It has become apparent that the international governance framework for food and agriculture of the last 50 years has failed to provide a sustainable food supply that is accessible to all. In the midst of this food price volatility, global leaders are seeking ways to reform the global governance of food in ways that will improve the global food system's ability to avoid future food crises.

Although the institutional infrastructure over the past half-century has generally succeeded in keeping food production abreast of demand, it has not been available to all, and an unconscionable number of nearly a billion people remain food insecure. Furthermore, recent experience highlights the fact that food supplies cannot be taken for granted and that national-level policy decisions, in addition to the functioning of the markets, can have destabilizing global impacts on prices and access to food well beyond their borders. Eliminating hunger is an area where we need to subordinate short-term national interests to broader global objectives. Reaching consensus on an appropriate framework to facilitate this is among the key challenges of international governance reform.

The need for a shared vision and agreement on global responses to problems that transcend national decisions led to the creation and subsequent evolution of the United Nations. The role of UN agencies in addressing global hunger remains a critical part of the food and agricultural governance reform agenda. This requires a balance between the interests of individual member country governments and common global-level objectives that are acted upon through UN bodies. Significant challenges within UN organizations include matching their mandates and means with the members' directives and the need to incorporate technical expertise and the evolving global knowledge base that is required to inform policy and program development.

The parallels between today's discussions and the dilemmas facing the founders of the international system are striking. The origins, experience in earlier food crises, and current reform of the UN Food and Agriculture Organization of the United Nations (FAO) provide a useful backdrop in examining these issues and can help to shed light on promising options for the future. The experience of FAO, and its place within the international food and agriculture governance system, highlights that we need to go forward on the basis of both vision and realism regarding the constraints to collective decisions and actions. This was true in 1945 when FAO was founded and is equally true now. FAO experience demonstrates the need for participation and buy-in by national governments and the recognition of the greatly expanded set of actors today. History since 1945 also illustrates the challenge of obtaining coherence in the overall system within a context of great swings up and down in the attention given to problems of agriculture and world hunger.

The Early Years: Parallels and Lessons

A striking similarity between 1945 and today is the vision of a hunger-free world and the difficulty reaching consensus on practical ways to achieve it. The concerns of those charged with a global response to food and agriculture problems of the postwar world were, as now, production, distribution, consumption, and trade, and the proposals called for an international architecture to deal with these elements in an integrated fashion. Behind this was a vision of food security as central to basic human dignity, economic development, and national and global security. The focus of the architects was not primarily on boosting production, although postwar reconstruction was, of course, a major concern, but rather on "freedom from want" and the connection between agriculture, hunger, and development. As the June 1945 General Report on The Work of FAO stated, "Freedom from want means the conquest of hunger and the attainment of the ordinary needs of a decent, self-respecting life.... If this can be done within and among nations by their separate and collective action, some of the world's worst economic ills, including hunger and extreme poverty will be on the way to extinction" (UN Interim Commission on Food and Agriculture 1945). That dealing with hunger was fundamental appeared obvious (Staples 2006, 86). The vision was clear; reaching consensus on how to go about it was not.

The European food situation was in crisis (Judt 2005, 21), and drastically reduced rice production in Asia added to the widespread malnutrition that had already existed before the war. FAO produced a World Food Survey in 1946 that highlighted the Great Depression's twin problems of food shortages and the renewed risk of unmarketable surpluses. In response, FAO's director-general, Sir John Boyd Orr, surveyed the existing organizations and found that a new global mechanism was required "to provide a means of acting together, as well as consulting together." He proposed a World Food Board to deal with production, distribution, and consumption (Staples 2006, 86).

The functions of the board would have been to (1) stabilize prices of agricultural commodities on world markets, including provision of necessary funds for these operations; (2) establish a world food reserve adequate for any emergency that might arise through failure of crops in any part of the world; (3) provide funds for the disposal of surplus agricultural products on special terms to countries where the need is most urgent; and (4) cooperate with organizations concerned with international credits for industrial and agricultural development and with trade and commodity policy (FAO 1946a).

The elements of the 1946 proposals, although not the single institutional mechanism, look remarkably familiar. The 2008 "Comprehensive Framework for Action" by the High-Level Task Force on the Global Food Security Crisis contains proposals on food assistance and social protection, increased investment to boost smallholder food production, and trade and international food markets (UN 2008). The 2008 G8 Declaration on Global Food Security echoed the need for "acting and consulting together" and proposed "a global partnership on agriculture and food, involving all relevant actors, including developing country governments, the private sector, civil society, donors, and international institutions" (ibid.). Expanding long-term investment in agricultural development and new mechanisms to ensure sufficient and timely food distribution are likewise central to the current debate on appropriate response to the food crisis and are the focus of several chapters in this volume. The earlier proposals for food reserves and a price stabilization mechanism are not so far in basic concept from the options put forward by Joachim von Braun and Maximo Torero of the International Food Policy Research Institute in their June 2008 paper "Physical and Virtual Global Food Reserves to Protect the Poor and Prevent Market Failure."

In 1946, these early proposals were taken up by FAO's member countries at the FAO Conference in Copenhagen. The Danish host called on the conferees to "let the world see that we are strong-hearted, far-sighted and wise enough to jointly lay what may become one of the great cornerstones of a sounder, happier and better world," but the comment by S.M. Bruce, the Australian representative (who had earlier experience with of the League of Nations on food and nutrition problems) that "the ideals of FAO were easier to state than to translate into actual accomplishment," was ultimately more accurate (Staples 2006, 90). The Conference unanimously adopted a resolution accepting the general objectives of the proposal and established a preparatory commission to submit recommendations regarding the necessary machinery (FAO 1946b).

The practicality of what would have been an international food marketing board, combining responsibilities on food distribution and credit for long-term agricultural growth was perhaps no greater in 1945 than today. The US State Department and the British Cabinet both objected to the specifics. The United States found the proposals "not only 'impracticable' but 'inimical to [America's] international trade policy'" (Staples 2006, 88). The United Kingdom, a large food importer at the time, was concerned about the potential rise in international food prices but sought a practical middle way.[1] Others saw trade issues as belonging to the (likewise stillborn) International Trade Organization,[2] or to the jurisdiction of the UN Economic and Social Council (ECOSOC) (ibid., 91–92).

More critical perhaps was the lack of conviction regarding the role of agriculture in development. The commission set up a joint committee on industrial development that argued that industrial development was in greater need of international cooperation than agriculture, which in any case already had sufficient funding. In the end, the preparatory commission recommended the creation of a World Food Council within the auspices of FAO and emphasized industrial development, full employment, and self help (ibid., 94). Attention on the food crisis waned,[3] and other global crises intervened at the start of the Cold War. In this era, plans for industrialization and import substitution as the best development path gained prominence. The preparatory commission's recommendations were never enacted, and at the next FAO Conference in Geneva, the members decided to keep FAO's budget at US$5 million and imposed a stricter program and budgetary review process that took some initiative away from the director-general and the secretariat (Staples 2006, 94).

In frustration, John Boyd Orr resigned from FAO in April 1948 but went on to receive the 1949 Nobel Peace Prize for his "great work in the service of mankind" and efforts to combat hunger. As Gunnar Jahn, chairman of the Nobel Committee, put in his presentation speech, "The World Food Board, which was to be invested with strong executive power, never became reality. It was too big a step to be taken all at once" (Jahn 1972). FAO, in turn, went on with a stronger emphasis on technical assistance and provision of what would now be called "global public goods and services," including early priorities of rinderpest eradication, locust control, rice technology, and international agriculture statistics. The experience and outcomes of the early deliberations shaped the evolving contours of both FAO and international governance of food and agriculture. Member countries shared a vision of the noble objective of combatting hunger and poverty but lacked consensus on how international institutions should play a practical role in this regard, particularly where this objective may impinge on trade or other national interests. They endorsed FAO's broad man- date and the need for an integrated approach to the crisis but did not agree on an integrated solution. Also evident were inevitable tensions between member countries and the secretariat, disagreement among members over priorities for the organization, and limited core resources to carry out what remained very ambitious objectives set by the membership.

The Evolving Context of Governance and Public Goods in Global Food and Agriculture

Although FAO was not accorded the powers that would have been vested in a food board, it was constituted with a mandate for the entirety of global food and agriculture. Indeed, except for the UN Security Council, none of the emerging architecture of the UN system was accorded such powers in areas of global governance as would have been vested in a food board. This is not to suggest that the FAO constitution did not place responsibilities on its members, in particular for disclosure of information on their food and agriculture situation, the basic building block for any moves toward global policy coherence. It was primarily for this reason that the former Soviet Union did not ratify its membership in FAO, although it had participated in the preparatory process.[4] The founding members similarly did not envisage FAO as the world's stand-alone organization on agriculture and food but rather as an organization that would be part of a knowledge network or partnership engaged jointly in the production, dissemination, and application of knowledge.[5]

For its first twenty years FAO held its wide mandate almost unchallenged. In 1962 FAO and the United Nations established the World Food Programme (WFP) to channel existing food surpluses in the developed world productively and to tackle emergencies. Then came the world food crisis of the 1970s, which had many parallels with the present price volatility. Food surpluses, especially in North America, had become taken for granted. In 1972, grain production fell simultaneously in many producing areas and imports rose, in some cases dramatically. Cereal prices more than tripled, and fertilizer prices more than quadrupled. FAO carried out most of the preparations for the 1974 World Food Conference designed to deal with this situation, but that conference also marked the first real questioning of FAO's capacity to address adequately all aspects of its mandate. This was a period during which response to issues was dominated by a project concept and a tendency to create new mechanisms or institutions rather than strengthen existing ones, a tendency also apparent in national development approaches where parallel activities of government departments became a norm.

The World Food Conference thus led to the founding of the International Fund for Agricultural Development (IFAD) as a specialized lender for pro-poor agricultural development. The UN General Assembly also established the World Food Council (WFC) to provide political leadership under the auspices of ECOSOC. The WFC, without technical capacity and paralleling the functions of FAO's governing bodies, was unable to develop this role and was suspended in 1993, an almost isolated case of suppression of a body in the UN system. With its demise, coherence on global policy and norm setting returned to FAO, with the significant exception of agricultural and food trade.

With regard to trade, as mentioned above, the establishment of an International Trade Organization failed, but in the parallel discussions for a trade treaty, the General Agreement on Tariffs and Trade (GATT) was agreed in 1947. GATT was not primarily concerned with agricultural products until the Uruguay Round of trade talks (1986–94), which also established the World Trade Organization (WTO). Following the Uruguay Round, global agreements on agricultural trade clearly became the prerogative of the WTO.

Agricultural research similarly evolved. The first four international agricultural research centres had been established largely by the Ford and Rockefeller foundations in the 1960s. These became federated into what is now the Consultative Group on International Agricultural Research (CGIAR) System in 1971. FAO was one of the co-sponsors along with the World Bank and the United Nations Development Programme (UNDP), but it rapidly became clear that the main initiative was with the World Bank and the donors. Technical agricultural research gradually slipped off the FAO agenda. Nevertheless, at the beginning of the 1980s, FAO was the biggest of the UN specialized agencies, with the largest technical cooperation program. This was also subject to change however, as UNDP made major changes in its policies for support to technical cooperation through national execution and WFP established itself as an autonomous program.

Finally, with the return of food surpluses, the realization of the importance of the social sectors in human development, and growing attention to the environment, agriculture slipped down the global agenda and with it, the perceived importance to the world of FAO's mandate. Along with the generalized decline in donor and national government allocations to agriculture, total resources available to FAO, not including emergency interventions, declined in real terms by 31 percent between 1994 and 2005.

The Current Reform and Renewal of FAO: Balancing Member Countries' Views, Expert Recommendations, and Management's Perspectives

It is against this background of a growing number of organizations addressing aspects of FAO's mandate, and a diminished FAO, that in November 2004 its member countries called for an independent external evaluation of the organization. All members valued some of FAO's services but almost all were also dissatisfied with its performance. Some of these members, especially among the developed countries, undoubtedly assumed that the evaluation would conclude on a further shrinking and concentration of activities in FAO. But by this time it was also becoming clear that problems of food and agriculture were again rising in importance on the international agenda, or at least were no longer in decline.

An immediate new world food crisis was not expected, but the underlying fundamental issues in producing enough food to satisfy needs were becoming very evident. These factors are now well known and discussed in more detail elsewhere in this volume.[6] At the same time the role the productive sectors must play in fighting poverty and, in particular, the extent of the dependence of the poor on agriculture was coming back into focus. FAO's mandate was back on the map, but the big questions for the evaluation were to be: What should be the role of FAO in fulfilling that mandate and what could be better done by others? How could FAO cost-effectively do the things that it was to do?

Although the first calls for the evaluation came from the developed countries, the evaluation was discussed in depth and its scope and organizational framework were designed with full participation by all the member countries through meetings of FAO's governing bodies. Members were also represented in large part by their "permanent representatives" to FAO in Rome. The planning process took almost a year, and in November 2005, the members approved the terms of reference for an Independent External Evaluation of FAO (the IEE) as the basis for reform.[7] This process led to a sense of ownership of the evaluation by the global membership, although within a framework that ensured the independence of the evaluation from both management (i.e., FAO's secretariat) and the governing bodies.[8] This was an evaluation called for, designed, driven, and owned by the membership of FAO. Although not imposed on the management, which welcomed the evaluation, neither was it co-owned by management, with negative as well as positive implications, as discussed below.

Several important aspects marked this evaluation as different from other reform panels, reviews, and evaluations in the UN system. It was the first to address the entirety of a major organization; all aspects of FAO were evaluated, not just the workings of its secretariat. In particular, the role and functioning of its governing bodies and FAO's role in achieving global policy coherence were evaluated. Second, it was designed to be an evidence-based formative evaluation, examining the emerging global context and, in the light of this, the needs, relevance, effectiveness, and efficiency of FAO, with a view to proposing solutions. It was not designed primarily to pass judgment on past performance but to draw on evidence-based findings to chart a road forward.

Equally important, the evaluation was not contracted to a consultancy firm, a move most members thought would have led inevitably to a bias in favour of Organization for Economic Co-operation and Development (OECD) countries, but rather was undertaken by a team selected on the basis of technical competence, with regional and gender balance. The evaluation was followed by a committee of the Governing Bodies, with tightly drawn terms of reference, as a guarantor of process and evaluation standards, but with no voice in the evaluation itself. Finally, the evaluation was adequately funded, with a budget of some US$8 million for the entire process, including preparation and immediate follow-up agreement on the recommendations.

The evaluation was conducted over a period of a year and a half and was divided into four main components: (1) The technical work of FAO; (2) management organization and administration; (3) the governance of FAO and FAO's role in global governance; and (4) FAO's role in the multilateral system. Standard evaluation tools were applied, including:

- review of existing evaluation and other evidence-based information on FAO's performance;
- preparation of an inception report for discussion, detailing how the evaluation would be implemented and with an initial identification of major issues for study;
- review of the academic literature and proposals made by a variety of stakeholders, including the non-governmental sectors;
- a set of country case studies;
- a series of surveys and technical papers by disciplinary specialists;
- comparisons with other organizations;
- benchmarking to best practice;
- statistically analyzable questionnaires to stakeholders; and
- participatory focus groups and individual interviews with stakeholders.

The independent external evaluation report "FAO: The Challenge of Renewal" stated its principal conclusion for FAO as *reform with growth*. It was quite critical: "The Organization is today in a financial and programme crisis that imperils the Organization's future in delivering essential services to the world....FAO's efforts are fragmented and its focus is on individual components of its vast challenge rather than the full picture.... The Organization has been conservative and slow to adapt.... FAO currently has a heavy and costly bureaucracy" (FAO 2007).

It went on to say that "the evaluation has concluded unequivocally that the world needs FAO and also that the problems affecting the Organization today can be solved." It found that FAO plays a unique role by combining the technical underpinning for informed and comprehensive global policy development in the areas of food, agriculture, forests, and fisheries and the forum for members to deliberate these policies. It further continued, "As a knowledge organization, FAO's job is to support Members in ensuring that the needs of the world in its areas of mandate are fully met—not necessarily to undertake each task itself" (ibid.). The evaluation concluded that FAO's role was essential to the global architecture and could be fulfilled by no other organization in areas including:

- keeping hunger, food, and agriculture on the global agenda as the world's attention flicks from one emerging crisis and quick fix to the next;
- developing global policy coherence and norms to address the issues;
- ensuring that the interests of the agricultural sector, the hungry, and the rural areas are not forgotten in global governance discussions from the environment to trade;
- information and statistics, including early warning, where FAO remained the most comprehensive source;
- drawing access to knowledge together for integrated application (technical, economic, social);
- facilitating coherence of action at country, sub-regional, and regional levels and continuing its technical advice and capacity building in areas of comparative advantage (technical areas were analyzed in some detail);
- mobilizing and facilitating coordinated action in food and agricultural emergencies; and
- identification of emerging issues.

The evaluation made over one hundred recommendations for institutional change to be incorporated into an immediate plan of action for FAO renewal. Upon receiving the report, the member countries established a committee of the whole to examine the recommendations and develop the action plan. This process led to a three-year Immediate Plan of Action (IPA) that converted the great majority of the IEE recommendations into actions and drew up an implementation budget. The plan was approved by a special session of the FAO Conference in November 2008, just twelve months from the receipt of the evaluation report by the FAO Conference in November 2007. It is regarded as the most far-reaching set of reforms to be agreed for any major multilateral organization.

Implementation of the Plan of Action has now passed to FAO management, with systematic reporting to the membership. A Reform Support Group has been set up to coordinate the implementation of the IPA and other more detailed aspects, including the "root and branch" review of management and administrative processes, a results-based management framework, and 360-degree performance evaluations. A Culture Change Team was also constituted with representatives from across the organization to help, as the FAO launching memo put it, "catalyse a process of culture change, identifying areas that need improvement through a truly interactive process, relying on the views and perspectives of staff at large throughout the Organization."

Reform has made a very positive start, but judgments on the impact or the extent to which reforms are implemented are premature and will follow at a later date. These reforms will depend on the internal and external capacity for change and on the budgets provided by the membership. FAO functions as a partner within the multilateral system, and parts of the renewal agenda depend not only on FAO but on other bodies as well. Other bodies will need to accept that FAO governing bodies and the secretariat have a legitimate voice on these global issues. So was it worth the cost? The response must be a clear yes. This was the first comprehensive evaluation and design of a renewal package in FAO's sixty-year history. The organization expends some US$1 billion per year, and even a marginal improvement in its effectiveness as a result of the process would be a very high return on investment.[9] Equally important, however, are the larger lessons and implications for institutional innovation to deal with the hunger crisis and challenges of food and agriculture in the twenty-first century.

Drawing Lessons from the FAO Experience

Lessons and Implications for Institutional Reform of the Component Parts of the System

The FAO experience illustrates the need for buy-in and ownership by national governments, and for consensus among members within the UN system, particularly between OECD members and the Group of 77 developing countries (G77). The FAO reform process was led by member countries, and this was essential to its ownership by national governments. This intensive involvement contributed to a positive spirit of working together for solutions across the OECD-G77 groupings seldom seen in the multilateral system. It also resulted in a commitment by members in 2007 for an FAO budget that for the first time in many years maintained purchasing power.

It was important that the FAO reform package was underpinned by an evaluation that was not only adequately resourced, professional, and independent but was *seen* to be these things. The greater the extent to which political decision making is informed by evidence-based analysis, the greater the possibility for agreement on solutions, and that those solutions will be both relevant and implementable.

It was critical that the external evaluation fed into a decision process agreed by member countries but that also had management involvement and acceptance. Important contributions to the formative review came from academic analysis and the ideas of non-governmental actors and experts, but these had to be weighed against political and institutional realities. The evaluation was an input to assist decision making by governments in the FAO Conference, not a substitute for that decision making. While management did not have a significant input in identifying issues, the evaluation itself recommended that the immediate plan of action should be developed jointly by management and the governing bodies. Although this did not happen immediately, the committee established by the FAO Conference gradually drew management into the process, and management became positively engaged.

From the beginning of the organization, there was recognition that food and agriculture problems cover a wide range of interrelated issues and require an integrated solution. A broad mandate for FAO is necessary; the dilemma both then and now has been to define realistic priorities and resources to carry them out. The FAO reform process illustrates how the competing views and interests of member countries, management, and external technical expertise can be combined. This is clearly important, but as the original architects of the system and the FAO evaluation recognized, it is only part of an integrated solution.

Implications and Lessons for Reform of the Overall System

Because of global attention given to the recent food crisis, there has perhaps never been a better time to make progress on change in international governance in order to reduce hunger and avoid repeated crises and repetition of what we have witnessed. Reform and adaptation of the components of the food and agriculture system are clearly not sufficient. As Mohammad S. Noori-Naeini, chairperson of the FAO Council, said in his presentation of the Immediate Plan of Action for FAO Renewal to the FAO Conference in November 2008:

> Many of the key players have now had comprehensive evaluations, including in addition to FAO, IFAD and the CGIAR system of international agricultural research centres....Most of the basic architecture is in place....But, despite the best efforts of all, the recent evaluations demonstrated that the international system, your international system, is not working as a coherent whole. The number of the world's poor and hungry continues to grow instead of decreasing in line with the World Food Summit and Millennium Development Goals.

> Thus my call to you now in the interests of all, but particularly those poor and hungry, is to make the necessary assessments and act swiftly to develop the coherence of the totality of the multilateral system to develop food and agriculture. (FAO 2008)

Achieving this change will require, as the 1945 architects of FAO put it, "separate and collective action" (UN Interim Committee on Food and Agriculture 1945). Effective policies and programs at the national level are essential and often missing, but achieving appropriate collective action remains a daunting challenge, particularly where this calls for subordination of short-term national interests to larger global objectives. The idea of a single institution in the form of a World Food Board was rejected in 1946 and never resuscitated. Nevertheless, the current moves toward a more integrated international architecture start from a promising base. The UN Secretary General's food crisis task force of senior officials in the UN system and Bretton Woods Institutions has completed its initial work, and there is progress by the Rome-based agencies in developing a common strategy (FAO, IFAD, and WFP). The CGIAR system is deeply engaged in its own reform process. Civil society organizations have also been very active in synthesizing diverse perspectives and getting their views on the table.

As the early experience illustrates, agreement by national governments on both the vision and practical measures to achieve it are required. Dealing with the longstanding hunger crisis will require that the concern expressed by many countries and stakeholder groups grows into a coherent demand from the majority of countries and their governments. It cannot be undertaken at the sole initiative of one group, for example, the G8, but may be catalyzed by such a group.

Strong and enlightened leadership is required, as well as sustained financial and political commitment. In addition to political leadership, however, it is clear that political will to tackle hunger is not often sustained spontaneously. The moral pressure for input into collective action that rises above the short-term national interests of the developed countries generally comes from civil society. Recent successes by the Jubilee Campaign of 2000 to diminish developing country debt and the movement to ban land mines show that this can be done. As expressed by other authors in this volume, the collective work and voices of myriad civil society organizations are an equally important part of the necessary international response framework.

Experience also indicates that further fragmentation of the system through the creation of additional institutions should probably be avoided. Instead, mandates, capacities, and mechanisms should appropriately be extended within the existing overall institutional architecture. Member countries will need to coordinate their positions internally and assert their vision for competing institutions, as the executive heads cannot be relied upon to spontaneously and rigorously integrate their work. Just as in the FAO evaluation, decisions on reform and strengthening of the international institutional architecture must be underpinned by an overall evidence-based analysis, drawing on existing evaluations of individual institutions and placing this in the global context. Such a balanced analysis will draw on ideas and evidence from all sources, confront the realities of each institution's current performance, and assist to catalyze and feed the political process by governments.

Just as in 1945, our world needs a coherent, relevant, and practical multilateral system and response to ensure that hunger will be abolished, that agriculture, forests, and fisheries will play their role in environmental sustainability, including addressing climate change, and that agriculture will make a contribution to lifting people out of poverty and ensuring economic growth. Just as in 1945, the world faces many competing challenges and there is a recognition that some solutions can only be obtained multilaterally. As in 1945, the world today recognizes that its hunger crisis requires sound international action. Unlike 1945, there is no Cold War developing, and although the current financial crisis and recession mean there are scarce resources, this is nothing like the scarcity or the devastation of national economies following World War II.

As in the 1970s, there is recognition that global governance and the provision of global public goods and services for food and agriculture need to be substantially strengthened. Unlike the 1970s, the emphasis is on enhancing the existing architecture and drawing existing institutions together, including a vastly expanded and active civil society and organizations engaged at the interface with food and agriculture in trade and the environment. We need to build on this rich experience, driving forward the political will to reach global objectives in this time of both challenge and opportunity. This is not an easy task but it is imperative.

Notes

1 According to Staples, the "British Cabinet sought a 'severely practical but positive and constructive middle way,' implementing those aspects of the proposals it thought beneficial and discarding Orr's 'extravagant and vague' formulations" (Staples 2006, 91–92).

2 A preparatory committee was established in 1946 to prepare the charter for an international trade organization, which was agreed in Havana in March 1948. The charter, however, was not ratified by key countries, and when the US government announced in 1950 that it would no longer seek ratification by Congress, the prospect of an ITO ended.

3 The food situation in Europe did not improve for several years, however, and Judt reports that "in French opinion polls taken in 1946, 'food,' 'bread,' 'meat' consistently outpaced everything else as the public's number one preoccupation" (Judt 2005, 86).

4 The Russian Federation as successor state to the USSR assumed its membership in FAO in 2006.

5 As the 1945 document put it, "Knowledge about better production methods, better processing and distribution, and better use of food is a first step.... How to get it put into practice on the necessary scale is the problem.... To surmount these difficulties will call for all the

wisdom and will that nations, acting by themselves as well as through FAO and other international organizations, can muster" (UN Interim Commission on Food and Agriculture 1945).

6 Some 1 billion people are badly malnourished today, and there will be a further 50 percent rise in the world's population by 2050. Added to this challenge are climate change, alternative uses of land and water, increased use of cereals for animal and fish production, more waste in food use with rising incomes, competition for agricultural products as an energy source, and a decline in the pace of productivity gains from technical innovation.

7 At the same time, FAO's governing body also approved an initial series of internal reforms proposed by the director-general.

8 Technical and administrative support to the process was provided by the FAO Evaluation Service, which has a dual line of reporting to the governing bodies and management, and functions with a high degree of independence.

9 The evaluation and design of the reform package had a direct cost of some US$ 8 million. The cost of the implementation of the three year Immediate Plan of Action is estimated as US$ 22 million for the first year of implementation. Cost savings will materialize in following years due to efficiency gains and savings in technical areas are planned to be immediately reapplied to technical work thus increasing impact but not appearing as a saving. If the total investment directly attributable to the renewal package was estimated to be US$ 50 million, this would still only constitute some 5 percent of FAO expenditure for a year.

Works Cited

Food and Agriculture Organization of the United Nations (1946a). "Proposals for a World Food Board." Washington, DC. 5 July.

——— (1946b). "Report of the Conference of the FAO, Second Session." Copenhagen. 2–13 September. http://www.fao.org/docrep/x5583E/x5583E00.htm.

——— (2007). "FAO: The Challenge of Renewal—Report of the Independent External Evaluation of the Food and Agriculture Organization of the United Nations (FAO)." C 2007/7A.1-Rev.1. Rome. September. ftp://ftp.fao.org/docrep/fao/meeting/012/k0827e02.pdf.

——— (2008). "Report of the Thirty-Fifth (Special) Session of the FAO Conference." C 2008/PV/2. Rome. 18–22 November. ftp://ftp.fao.org/unfao/bodies/conf/c2008/PV201.doc.

Jahn, Gunnar (1972). "The Nobel Peace Price 1949: Presentation Speech." In *Nobel Lectures, Peace 1926–1950*, ed. Frederick W. Haberman. Amsterdam: Elsevier.

Judt, Tony (2005). *Postwar: A History of Europe Since 1945*. New York: Penguin Press.

Staples, Amy L. (2006). *The Birth of Development: How the World Bank, Food and Agriculture Organization, and World Health Organization Changed the World, 1945–1965*. Kent, OH: Kent State University Press.

United Nations Interim Commission on Food and Agriculture (1945). "The Work of FAO: A General Report on the First Session of the Conference of the Food and Agriculture Organization of the United Nations." Washington, DC.

United Nations, High-level Task Force on the Global Food Security Crisis (2008). "Comprehensive Framework for Action." New York. http://www.un.org/issues/food/taskforce/Documentation/CFA%20Web.pdf.

von Braun, Joachim, and Maximo Torero (2008). "Physical and Virtual Global Food Reserves to Protect the Poor and Prevent Market Failure." IFPRI Policy Brief 4. Washington, DC: International Food Policy Research Institute.

CHAPTER 14

Improving the Effectiveness of US Assistance in Transforming the Food Security Outlook in Sub-Saharan Africa

Emmy Simmons and Julie Howard

The global food-price crisis of 2007–08 brought home the dramatic implications of two decades of world neglect of agricultural development. Consumers in many developing countries suddenly found themselves unable to find—or afford—the foods they needed. Spiking prices sparked instability and civil unrest in nearly forty countries in early 2008. Many of the outbreaks were in sub-Saharan Africa, a region that, since the 1980s, has increased its imports of foodstuffs significantly, responding both to the demands of a growing population and lagging regional agricultural growth (World Bank 2008a; World Bank 2008b). While agricultural productivity has risen in several countries since 2000, increases have not been high enough to reduce imports or to improve food security substantially. The Food and Agriculture Organization of the United Nations (FAO) estimated in 1996 that some 186 million people in sub-Saharan Africa did not consume enough calories to provide the energy needed to live an active and healthy life. In 2008, the comparable figure was 263 million (FAO 2008).

The US and Other Donors Have Emphasized Humanitarian Responses

Humanitarian assistance—principally food aid—has been an important short-term donor response to Africa's rising food insecurity. Africa's share of total humanitarian assistance provided by members of the Organization for Economic Co-operation and Development (OECD) grew from 31 percent in

1995 to 46 percent in 2006 (Development Initiatives 2008). The volume of such assistance rose every year between 1998 and 2005. Food aid is the largest single component of humanitarian assistance, and sub-Saharan Africa is the primary recipient (OECD 2009). In 2007, the World Food Programme (WFP) directed more than 72 percent of total WFP food assistance (worth approximately US$2.2 billion) to the region (WFP 2008). The United States remains a key source of food aid, having provided nearly half of all global supplies through both multilateral and bilateral channels since 1990 (WFP 2009). US funding for food aid has averaged approximately US$2 billion per year since 2001. In 2007, 74 percent of US Title II emergency food aid and 47 percent of Title II non-emergency food aid resources were allocated to sub-Saharan Africa (USAID 2008).

Meanwhile, donor investments in agricultural development have declined. From 1990 to 2007, bilateral aid programs directed fewer than 5 percent of their assistance resources to the agricultural sector (OECD 2009). The trend has been strongly downward, falling to less than 3 percent of total official development assistance in 2006. Traditional US assistance for agricultural development in sub-Saharan Africa from the US Agency for International Development (USAID) averaged just US$324 million annually during 2000–04. When assistance from other US government programs and US contributions to multilateral organizations are included, the total rises, but remains less than US$500 million per year (Taylor and Howard 2005).

Total US bilateral agricultural development assistance for sub-Saharan Africa increased from an estimated US$408 million in 2004 to an estimated US$720 million in 2008, but the gain in bilateral assistance was due almost entirely to commitments made by the US Millennium Challenge Corporation (MCC) (Taylor 2009). The MCC was first established in 2004, but by 2008 it had committed more than US$1.8 billion to agriculture-related investments in nine countries in sub-Saharan Africa, specifically Benin, Burkina Faso, Cape Verde, Ghana, Madagascar, Mali, Mozambique, Namibia, and Tanzania (MCC 2009). These commitments were shaped in response to proposals from African governments and focused heavily on agriculture, agriculture-related infrastructure, and other economic growth investments. However, the outlook for future investment is uncertain. The US Congress has reduced the MCC budget over the last two years, and further MCC commitments have been placed on hold.

There is increasing consensus in the United States that the balance between development and food assistance must change. Food aid is still needed to meet the needs of the chronically poor and those whose livelihoods have been devastated by conflict or natural disasters. But without greater investment in Africa's own capacities to produce and market the commodities needed to feed a growing population, and to generate the kind of pro-poor economic growth that can only come from agricultural development, the risk of repeated emergencies is likely to rise and chronic poverty will persist. The impact of the current global economic downturn is being rapidly transmitted to Africa and is reflected in lowered projections for growth in 2009–10. Investing more in agriculture—an area of comparative advantage—will help many African countries to weather the recession, regain a pathway to economic growth, and improve their food security outlook.

The World Bank and other bilateral and multilateral donors are already moving toward more substantial investments in agriculture. Private foundations such as the Bill and Melinda Gates Foundation have also become important partners in the revitalization of African agriculture, with the foundation's flagship Alliance for a Green Revolution in Africa (AGRA) providing both financial support and strong advocacy. Therefore, the US government will find there are willing partners with whom to collaborate. The financial crisis may place commitments of additional assistance and local investments in jeopardy, but the emerging reality is that all partners—in Africa and in the international community—will have to do "more" and "better" with limited new funding to meet the twin challenges of food security and agricultural development.

Africa's Commitment to Agriculture

In 2003, African heads of state issued the "Maputo Declaration" on Agriculture and Food Security. This declaration called for the urgent implementation of the Comprehensive Africa Agricultural Development Programme (CAADP) developed under the aegis of the African Union's New Partnership for Africa's Development (NEPAD). CAADP provides a framework for cooperation among the nations in the region, defining four priority areas—or "pillars"—for action:

1. Land and water management, which aims to extend the area under sustainable land management and reliable water control systems.
2. Market access, focusing on improved rural infrastructure and other trade-related interventions.
3. Food supply and hunger, which emphasizes raising smallholder productivity and improving responses to food emergencies.
4. Agricultural research, including the strengthening of systems needed to disseminate appropriate new technologies (CAADP 2009).

The declaration further committed the signatory governments to boosting the priority of agriculture in their national budgets, allocating at least 10 percent of national resources within five years. Steady progress has been made since then.

In several countries, agricultural growth has matched or exceeded the target of 6 percent per year, and Mali, Madagascar, Malawi, Namibia, Niger, Chad, and Ethiopia have met the 10 percent investment target (Mkandawire 2009). Regional economic communities such as the Common Market for East and Southern Africa (COMESA) and the Economic Community of West African States (ECOWAS) have focused on the development of regional market infrastructure and agricultural trade. CAADP is facilitating structured country-level discussions in order to frame agricultural development priorities for African countries and their donor partners.

External Assistance, African Capacity, and Greater Impact

While it is critical that African governments take steps to reach the CAADP targets, it is also clear that external resources—both public and private—will be important for realizing the scale of investments needed to have a major impact on food security and agricultural development in sub-Saharan Africa. Estimates of public and private financing needs for expanding irrigation, transportation infrastructure, and energy access have been recently developed in the Africa Infrastructure Country Diagnostic (AICD) project led by the World Bank. Given the limited reach and poor quality of current African infrastructure, these estimates imply a major increase in public financing over the next decade—from the current US$40 billion per year to US$80 billion annually (AICD 2009). Similarly, the International Food Policy Research Institute has developed a strategy for rapid expansion of agricultural research that focuses on "best bets" in research; the annual cost of this program is estimated to be between US$4.6 and US$9.3 billion in public investments (von Braun et al. 2008).

In responding to these proposals, multilateral development assistance organizations and bilateral organizations, such as USAID and the MCC, need to consider not only *what* is funded but *how* it is funded. Rather than substitute for African initiative, effective external assistance must support and extend it. Partnerships between internal and external actors are needed, for example, to improve governance affecting food security and agricultural development, spur the development of private African agricultural enterprises in addition to foreign investment in agribusiness, and ensure that African civil society can effectively engage in decision making that affects food security and agriculture and rural development.

The United States has a long history of successful assistance for institution-building in Africa, although as agricultural budgets have declined, this history has become tarnished. Expanded US investments to strengthen the capacity of those institutions and organizations responsible for developing and sustaining the process of agricultural growth could yield immediate benefits in terms of greater ability to manage increased resource flows but, more importantly, could build the foundation for long-term sustainable agricultural growth. This implies greater support for national and regional institutions and organizations that:

- provide public goods: higher education, research, regulation, and support of cooperatives and other farmer organizations, market information systems, and infrastructure;
- support private-sector investment: banks and microfinance organizations, agribusiness firms that provide inputs and processing services, trading operations, regulatory agencies, labs and other certification organizations; and
- foster grassroots awareness of issues and enable diverse groups, including those that represent women's interests, to advocate for food security and agricultural growth.

It is impossible to ignore the negative impact of the increasing variance and volatility of prices and policies on the agricultural investment decisions of individuals and companies. To counter this, African governments, the United States, and other partners must focus much more on reducing risk and boosting resilience in the agriculture sector. Protectionist instincts—and policies—have emerged in countries seeking to buffer their citizens against global turmoil. Exports of staple commodities have been banned, consumer prices have been fixed at levels that do not provide adequate returns to farmers, and governments have expanded subsidies to boost production even as tax revenues have dropped. However, there is broad agreement that a return to the self-sufficiency strategies prevalent in Africa during the 1970s and 1980s will not be effective. Rather, it is likely to cut poor farmers and consumers off from regional and global trade and leave them even more susceptible to the risk of food insecurity.

The Way Forward: Principles and Options

African and American leaders agree that expanded, long-term support for agricultural development is essential if African countries are to build sustainable pathways out of poverty and meet the needs of the more than 263 million people who are hungry and food-insecure. To paraphrase the new US president, the United States, Africa, other donors, the private sector, and civil society must work alongside each other to make African farms flourish.

This will require some changes in approach. Former USAID Administrator Peter McPherson noted in opening remarks to the 2009 "US–African Forum on Improving the Effectiveness of US Assistance and Investments in Challenging Economic Times" that the MCC had demonstrated the value of allowing programs to be country-driven rather than donor-driven. Data show that MCC-eligible countries have proposed substantial investments in agriculture and rural development. The collaborative process of negotiating MCC "compacts" has also contributed to greater flexibility in program design. However, these innovations have also led to a two-tiered US assistance program: African countries that qualify for MCC are able to set their priorities and discuss them in detail with the funding organization, while those countries that cannot qualify for MCC funding take what they can get from a number of other US assistance programs. "This approach needs to change," McPherson said. "All recipient countries need to have input into US government investments."

Specific suggestions for improving the effectiveness of US assistance/investments in African agriculture and food security emerged from broad discussion among the more than 140 participants in the forum. Although consensus was not reached on every point, several proposals gained considerable support and merit further attention. The following suggestions for US policy are among the ideas put on the table at the forum.

Coordinated Approach

Adopt a coordinated, whole-of-government approach to plan and monitor US support and policies related to African food security and agricultural development. In addition to USAID and MCC, such an approach would require integration of support from the Department of State Office of the US Trade Representative, the Treasury Department, the US Department of Agriculture, and the US President's Emergency Plan for AIDS Relief (PEPFAR). Further, US programs should be "balanced"; that is, capable of meeting emergency needs, but reoriented, significantly increasing US investments in agricultural development to assure sustained food security over the longer term. In Africa, the US government should create "one-stop shops" for development assistance, and use annual consultations featuring an integrated team of US agency representatives empowered to prepare, monitor, and adjust country- and region-specific development assistance and related trade, health, and nutrition policies and programs.

The African Agricultural Agenda

Embrace the African agricultural agenda outlined in the CAADP and commit to active collaboration with bilateral and multilateral partners in countries and regions around this African-defined agenda, measured against mutually defined and monitored objectives and target outcomes. This also implies that the United States should follow the African lead in recognizing that accelerating Africa's agricultural and economic growth will depend on effective regional integration of markets, trade, and supporting institutions. The overwhelming majority of US assistance to the region is country-based. The United States should give greater priority to the development of regional, not just bilateral, investments to support regional economic integration.

Strategic Decisions

Make strategic decisions about where the US can be most effective at national and regional levels, based on US strengths, deep knowledge of country and regional priorities, of other partners' programs, and the pace of progress toward mutually defined objectives. Potential reforms in four areas widely believed to be areas of US strength were addressed in some detail.

Food aid

More effective US food assistance programs would enable greater flexibility in design, so that the United States can deliver the most appropriate support, whether commodities from the United States or from local or regional sources, vouchers, or cash-based programming for agriculture and food security. Cost-effectiveness and timeliness should be important considerations. The nutritional quality of food aid provided through US programs must be improved and adapted to the specific nutritional needs of recipients according to the best scientific guidance. Local and regional purchase programs should be monitored and expanded in ways that promote small-scale production and market development.

Capacity-building

There is a need to strengthen organizational capabilities at all levels (regional economic communities, national government agencies, farmer organizations, and commodity groups at national and regional levels). Building education institutions, especially higher education and secondary technical schools, is a particular area of emphasis for many African countries. Strengthening science and technology innovation and delivery for agriculture, by linking research, extension, and education institutions, is an area of clear comparative advantage for US assistance.

Business solutions for agricultural development and food security

The United States should consider scaling up promising business solutions by helping to strengthen the policy and regulatory environment and ensuring the level and quality of public-sector investments needed to attract more and better funding for small and large agriculture projects. Enhancing private-sector entrepreneurial capacity has been a key area of program interest for the United States under the African Growth and Opportunity Act (AGOA), but progress in the agricultural sector to date has been minimal. Reframing AGOA to include development of regional trade opportunities as well as Africa–US trade could provide a new platform for collaboration.

Expand economic infrastructure

Strengthening transportation, communication, and water and power networks is vital to raising agricultural productivity, improving markets, and creating a conducive environment for private-sector investment. US assistance could help develop criteria for the allocation of scarce investment funds at regional and national levels and involve the private sector and civil society in an informed decision-making process. Infrastructure needs on the continent are so great that stronger US collaboration and integrated planning of infrastructure programs with country, donor, and multilateral agency partners is imperative.

Modify Management Structures and Approaches

It was suggested that modifying the management structures and approaches that govern US assistance would provide incentives for the establishment of long-term goals and objectives that transcend the terms of individual US program leaders. These changes would recognize that progress in strategic areas often requires solid commitments beyond the typical two-to-four year project cycle. New management approaches should also create positive incentives for collaboration and co-investment on major projects with local, bilateral, and multilateral partners. They would also move decisively toward knowledge-based development programming by increasing the transparency of aid flows and impacts, significantly expanding funding for independent monitoring and evaluation of projects, and developing, in collaboration with donor, multilateral, and African partners, credible indicators for the quantity and quality of agricultural development investments by governments and their donor partners.

Remaining Questions

The depth of the proposals tabled in the US–African Forum discussions reflected the rich experience of the participants. But several questions emerged in debate for which answers were not readily apparent. These questions form the core of a new agenda for analysis and consideration at both bilateral and multilateral levels, in donor capitals, and in African countries.

The availability of private-sector *financing and investment capital* in sub-Saharan Africa is likely to be affected by the global economic downturn. Public financing will not compensate. What are the priorities for investment, and what steps might the US government or other donors take to increase the confidence of private investors regarding opportunities in African agriculture and supporting infrastructure? Could additional capital be leveraged through public–private partnerships? How can warehouse receipts and other innovations be institutionalized to help many smallholders and entrepreneurs gain access to short- and long-term capital from banks in the absence of traditional collateral such as land titles? What other market-stabilization tools might be helpful? Could virtual and physical grain stocks being proposed at global, regional, and national levels avoid the pitfalls of physical stock systems prevalent in sub-Saharan Africa during the 1980s and 1990s? What is the role of input subsidy programs?

How might the *lessons of success* in agricultural development elsewhere in the world be better adapted or revised for African use? Many tested and successful US approaches to agricultural development, for example, would seem to be useful to Africa; for instance, land-grant universities that bring together research and education, cooperative farmer organizations and credit associations, warehouse receipts, and so on. Some of these approaches have worked where they have been introduced; others have not. How might lessons from Asia and Latin America be teased out to serve as useful guidance for Africa's own Green Revolution?

There is broad agreement that *science and technology* have an important role to play in increasing productivity in African agriculture. But it is not clear what the balance of technology-related investments should be. How much should be allocated for basic science, improving understanding of emerging pests and diseases, for example, relative to the investments made to extend technologies that meet farmers' needs? Do we have the right organizations to do both research and technology transfer in Africa? Is there a need for broader partnerships?

We do not have an agreed set of *metrics* to measure the quality of development assistance for agriculture and food security. Program effectiveness means different things to different people. To some, it means better processes: closer working relationships, programs reaching more farmers, or farm-level interventions reducing risks. To others, it means impact: sustained access to adequate supplies of nutritious food, women accessing farmland and extension services, or higher yields per hectare. Still others focus on indicators of program performance: percentage of national budgets allocated to agriculture, economic returns per dollar of assistance, numbers of students trained and employed, or inclusion of vulnerable groups. When assistance or investment resources are limited, it is important not to over-promise on expected results.

Conclusion

Never before has the divide between the world's rich and poor been more glaring. The problems are particularly acute in sub-Saharan Africa, where nearly half of the region's population lives on less than one dollar a day, and more than a third lack basic food security. And sub-Saharan Africa's conditions are deteriorating: it is the only region of the world where poverty and hunger are projected to increase over the coming decades.

Agricultural development is a critical catalyst for economic growth and poverty reduction in sub-Saharan Africa. Three-quarters of the population live and work in rural areas. GDP growth in agriculture has large potential benefits for the poor and is at least twice as effective in reducing poverty as growth generated by other sectors (World Bank 2007).

African leaders have recognized this reality and have committed to greater public investments, expanded regional trade, and more collaborative action within the framework of the CAADP. Donors, including the United States, have begun to reorient their assistance to support these efforts, especially in response to the global food price crisis of 2007–08. A more substantial and collaborative effort is needed. There are many areas in which more public funding is essential—to invest in infrastructure, research, extension services, higher education, and the like—but there are also many areas where a change of approach could make those resources that are available more effective. The United States is a critical partner in Africa's development and must do much more to accelerate agricultural growth and improve the outlook for regional food security, even in these challenging economic times.

Note

This chapter draws on discussions held at a February 2009 conference convened by the Partnership to Cut Hunger and Poverty in Africa in Washington, DC: "Transforming Food Security and Agricultural Development in Sub-Saharan Africa: A US–African Forum on Improving the Effectiveness of US Assistance and Investments in Challenging Economic Times." During two days of plenary and small-group discussions, participants considered how the US can work more effectively with its partners to address the urgent imperative of improving short- and long-term food security in sub-Saharan Africa. The goal of the forum was to identify key problems and specific, practical approaches to improve the impact of US assistance and investments in Africa.

Works Cited

Africa Infrastructure Country Diagnostic (2009). "Africa's Infrastructure: A Time for Transformation. Summary of Findings." http://www.ppiaf.org/documents/FINAL_AICD_Brochure_English.pdf.

Comprehensive Africa Agriculture Development Programme (2009). "The Comprehensive Africa Agriculture Development Programme." Midrand, South Africa. www.caadp.net.

Development Initiatives (2008). "Good Humanitarian Assistance, 2007/2008." Somerset. http://www.goodhumanitariandonorship.org/documents/gha_2007_final_a4.pdf.

Food and Agriculture Organization of the United Nations (2008). "The State of Food Insecurity in the World, 2008." Rome. ftp://ftp.fao.org/docrep/fao/011/i0291e/i0291e00.pdf.

McPherson, Peter (2009). Opening remarks delivered at US–African Forum on Improving the Effectiveness of US Assistance and Investments in Challenging Economic Times: Transforming Food Security and Agricultural Development in Sub-Saharan Africa. Washington, DC: Partnership to Cut Hunger and Poverty in Africa. 23 February.

Millennium Challenge Corporation. (2009). "MCC Investments Contribute to Long-Term Food Security." Fact Sheet. Washington, DC. 20 March. http://www.mcc.gov/documents/factsheet-032009-foodsecurity.pdf.

Mkandawire, Richard (2009). "How the CAADP Process Can Accelerate Agricultural Development and Improve Food Security in Africa over the Next Decade." Presented at US–African Forum on Improving the Effectiveness of US Assistance and Investments in Challenging Economic Times: Transforming Food Security and Agricultural Development in Sub-Saharan Africa. Washington, DC: Partnership to Cut Hunger and Poverty in Africa. 23 February.

Organization for Economic Co-operation and Development (2009). "OECD.Stat Extract." Paris. http://stats.oecd.org/WBOS/index.aspx.

Taylor, Michael R. (2009). "U.S. Agriculture-Related Development Assistance for Sub-Saharan Africa: Key Trends." Washington, DC: Partnership to Cut Hunger and Poverty in Africa Issue Update.

Taylor, Michael R., and Julie A. Howard (2005). "Investing in Africa's Future: U.S. Agricultural Development Assistance for Sub-Saharan Africa." Washington, DC: Resources for the Future and Partnership to Cut Hunger and Poverty in Africa. http://www.africanhunger.org/uploads/articles/ab119510183f8e254629783f67ea6abe .pdf.

United States Agency for International Development (2008). "U.S. International Food Assistance Report 2007." Washington, DC. http://www.usaid.gov/our_ work/humanitarian_assistance/ffp/fy07_usifar_final.2008.pdf.

von Braun, Joachim, Shenggen Fan, Ruth Meinzen-Dick, Mark Rosegrant, and Alejandro Nin-Pratt (2008). "International Agricultural Research for Food Security, Poverty Reduction, and the Environment: What to Expect from Scaling Up CGIAR Investments and 'Best Bet' Programs." IFPRI Issue Brief 53. Washington, DC: International Food Policy Research Institute. http://www.ifpri.org/pubs/ib53.pdf.

World Bank (2007). "World Development Report 2008: Agriculture for Development." Washington, DC. http://siteresources.worldbank.org/INTWDR2008/Resources/WDR_00_book.pdf.

——— (2008a). "Food Crisis." Washington, DC. 30 March. www.worldbank.org/html/extdr/foodprices/.

——— (2008b). "Framework Document for a Global Food Price Crisis Response Program." Washington, DC. 26 June.

World Food Programme (2008). "Annual Report 2007." Rome. http://www.wfp.org/sites/default/files/2007_Ann_Rep_English_0.pdf.

——— (2009). "World Food Programme." Rome. http://www.wfp.org/.

CHAPTER 15

Urban Agriculture and Changing Food Markets

Mark Redwood

In the spring of 2008, rapid increases in the price of basic foods galvanized media attention across the world. While the speed of these changes seemed to catch many off guard, an argument can be made that rising food prices and the resultant "crisis" have been a long time coming. What is clear is that volatility in world food markets is having a serious impact on the world's poor. The end result is rapidly deteriorating food security for those without reliable access to basic foods. As a result of this, an estimated additional 290 million people are at risk of falling into poverty because they cannot service basic household food needs due to the high costs involved (Oxfam 2008). Table 15.1 further illustrates the dramatic changes seen in real price increases for rice, a staple food, in several countries.

As discussed in earlier chapters, price increases in food markets are a result of a combination of supply and demand factors that temporarily displace price equilibrium. Supply and demand factors are affected by dynamic change resulting from cyclical market factors, economic/political structures, or as a result of the changing global environment. Pro-poor subsidies have insulated the poor from some of the more extreme fluctuations in prices (see Table 15.1). However, the cost is then simply borne by the public purse. Still, developing countries have little influence over the economic drivers that have pushed prices higher.

It is clear that volatile prices are a threat to food security worldwide, but for those with limited access to food and markets, the threat is even greater. The definition of food security has undergone many iterations, but one of the more recent versions resonates well: "Food security [is] a situation that exists when

Table 15.1
Cumulative Percentage Changes in Real Rice Prices, Fourth Quarter of 2003 to Fourth Quarter of 2007

Country	(1) World price (US$)	(2) World price (DC)	(3) Domestic price (DC)	(4) Pass through (%) = (3)/(1)
Bangladesh	56	55	24	43
China	48	34	30	64
India	56	25	5	9
Indonesia	56	36	23	41
Philippines	56	10	3	6
Thailand	56	30	30	53
Vietnam	39	25	3	11

Column 3 represents the real percentage increase in Domestic Currency (DC). Where the number is low, countries have subsidized or regulated the market in favour of moderate prices.

Source: Dawe (2008)

all people, at all times, have physical, social and economic access to sufficient, safe and nutritious food that meets their dietary needs and food preferences for an active and healthy life" (FAO 2002). Current volatility is now compromising, for many of the poor, economic access to sufficient food. Moreover, the problem of food security is an increasingly urban one, as those living in cities are exposed to market volatility without having access to land to grow food for household consumption.

The Rise of Cities and Changing Food Consumption Patterns

Possibly the most significant migration in human history has been the transfer of people from rural areas to cities and towns. The oft-cited statistic from the United Nations notes that in 2007 more than 50 percent of humanity was living in cities and towns. This demographic shift reflects the importance of the economic power of cities. Much urban growth stems from population growth within cities (60 percent), but another 30 percent is attributed to rural to urban migration. Most annual growth rates for cities in developing regions of the world are between an astounding 4 and 6 percent (Tannerfeldt and Ljung 2006). The implications of this change can be dramatic from a food security standpoint.

A rapidly urbanizing world has serious implications for food availability. Urbanization is associated with economic growth and development, which involve changing diets toward more energy-intensive protein foods (i.e., meat). This food requires more land per kilojoule (kJ) produced. For instance, while the human population has grown at a rate of 1.1 percent per year between 1996 and 2005, pig and poultry production has grown 2.6 and 3.7 percent respectively during the same period (Otte et al. 2007). Meanwhile, given the history of settlement patterns, cities are often located on the best agricultural land. As cities grow, this land is converted for urban development. Furthermore, food in cities is often processed and urban dwellers rely less on staples, which are generally more nutritious.

A number of factors impact food security in cities. For example, the decline in food security is related to the integration of food markets internationally (Koc et al. 1999). Another factor is the rural to urban migration that has affected rural food-producing regions and increased the number of people in cities reliant on others to provide their food. Frayne studied food security in Namibia and found that most migrants were heavily reliant on food transfers from family and social networks in rural areas, suggesting the existence of complex social networks that help to reduce the food uncertainty associated with reliance on markets (Frayne 2005). Other factors include the removal of food subsidies, leading to more exposure for the poor to volatile prices, and the fact that high-quality, locally produced foods are often export crops with limited local availability.

The overall result is that, in many cities, particularly those that are industrializing, poor residents face food insecurity; that is, a lack of consistent access to healthy and nutritious food (Dixon et al. 2007). The problem is particularly acute among those poor whose reliance on market sources of food is not matched by adequate wages with which to buy that food. With only tenuous rights to land, or in many cases no access at all, this presents a serious dilemma.

Food Prices and the Urban Poor: The Spectre of a Major Crisis

The Food and Agriculture Organization (FAO) points out that the urban poor are disproportionately affected by rising food prices. There are two main reasons offered for this. First of all, city dwellers are more likely to consume foods that are tradable commodities (e.g., wheat or rice) and thus are more exposed to market volatility. Conversely, in rural areas, diets are often made up of traditional staples such as roots and tubers. Secondly, city residents have much less access to land and other inputs required to grow their own food (FAO 2008). This naturally increases their exposure to fluctuating prices and leaves them with few options to react to changing prices.

High food prices can also impact the poor by exacerbating problems that already affect the household. A striking example is apparent in Southern Africa where some research finds that high food prices are linked to an increase in prostitution and HIV exposure (IDRC 2008). This work also confirms the perhaps logical notion that increases in food prices reduce the resources available to care for the ill and to support vulnerable children. This food insecurity is now placing an increased burden on city residents whose employment fuels remittances to their families in rural areas.

Urban dwellers also tend to spend a large proportion of their income on food. Table 15.2 illustrates the extent of this with research finding that, in some cities, the poor spend an astounding 60 to 85 percent of their income on food. Clearly, the higher the proportion of household income spent on food, the more vulnerable its residents will be to sudden or extreme changes in price.

The run-up in prices has also had a destabilizing impact on some cities whose vulnerable residents have been adversely impacted. Dakar and Cairo are two cities that have witnessed social unrest directly associated with the impact of volatility in food commodity markets. In March 2008, up to nine people were killed in Cairo in bread riots associated with high commodity prices and the reduction of subsidies (Kliger 2008). In Dakar, Senegal—a country that imports 80 percent of its rice—high food prices and a crackdown on informal sellers led to several protests and much social unrest in late 2007 and early 2008 (Ba 2008).

Table 15.2

Percentage of Income Spent on Food by Low-Income Residents in Selected Cities

City	Income spent on food (%)
Bangkok (Thailand)	60
La Florida (Chile)	50
Nairobi (Kenya)	40–60
Dar es Salaam (Tanzania)	85
Kinshasa (Congo)	60
Bamako (Mali)	32–64
Urban USA	9–15

Source: Akinbamijo, Fall, and Smith (2002)

Urban Agriculture: Coping with Crisis

Urban agriculture (UA) is one commonly employed strategy to reduce the impact of fluctuating commodity markets on the poor. The classic and widely used definition of UA comes from Mougeot:

> Urban Agriculture is an industry located within, or on the fringe of a town, a city or a metropolis, which grows and raises, processes and distributes a diversity of food and non-food products, (re)using largely human and material resources, products and services found in and around that urban area, and in turn supplying human and materials resources, products and services largely to that urban area. (Mougeot 2000)

This definition links confined space production, interrelated economic activity, location, destination markets (or home consumption), and the types of products produced in a dynamic interaction that can vary from one urban area to another. The breadth of this definition has not been challenged, and it influences the extent of research on the subject.

Urban agriculture also acts as a catalyst for political organization. A survey of producer groups from 2005 to 2007 identified that organizations based around UA play a significant role in social cohesion, offering technical training and providing a platform for political lobbying. Such groups have successfully lobbied for municipal policy change (Amsterdam); acted as the voice of farmers to lobby for formal recognition (Dakar, Villa Maria Triunfo); or have been provided technical assistance (Montreal) (Santanderau and Castro 2007).

The administrative limits of a city also play a role in determining the extent of the practice. Work in Latin America, supported by the International Development Research Centre (IDRC), demonstrated that within the limits of the cities in the region there were large areas of vacant land. For instance, in Quito, Ecuador, 35 percent of city land was vacant and often used for agriculture (2001 data). In Rosario, Argentina (2003 data), the amount was 80 percent (IDRC 2004). Recent data from Abomey and Bohicon, two cities located in Benin, West Africa, shows that agriculture is the main activity for 3 to 7 percent of people living in the downtown core. However, six kilometres from the city limits, in the peri-urban area, the percentage grows to 50 (Floquet, Mongo, and Nansi 2005). In the five urban districts that make up downtown Hanoi, 17.7 percent of land is used for agriculture (Mubarik et al. 2005).

The status quo response to UA by governments tends to be prohibitive of the practice. Such policy stems from a perception of UA as a form of resistance to urban development priorities as determined by planners. Some cities have, by virtue of being exposed to UA and farmer groups, changed their perspective and put in place systems that are designed to support UA, or at least remove the most draconian restrictions on the activity. However, even when rules are in place, they are often not well understood or enforced. In Harare, Zimbabwe, 40 percent of UA practitioners were unfamiliar with any laws related to it (Redwood 2008). Moreover, one in five considered the existing legislation to be hostile toward the practice.

Nonetheless, progress is being made. The number of municipalities that have policies in favour of UA has increased dramatically in recent years. Accra, Beijing, Brasilia, Buluwayo, Governador Valdares, Havana, Hyderabad, Kampala, Rosario, and Nairobi are a short list of the growing number of cities that have made significant progress in this area. Another popular method of supporting urban farming has been the establishment of food-policy councils. These councils represent an increasingly common way of bridging community groups with municipal politicians and bureaucrats. Amsterdam, Toronto, Vancouver, London, Detroit, and Pittsburgh all have councils that encourage locally based food systems.

The growing research on UA, and the scale of it, suggests its importance and points clearly to the role it may play in helping to adapt to the new realities of food markets. Table 15.3 identifies several pieces of data on the percentage contribution of UA to urban food availability.

By its very nature and its geography, UA is a system that prioritizes local markets, involves very little transportation since it operates directly next to its main market, employs both liquid and solid waste (re)use, and is an economic activity that employs not only farmers but others engaged in associated support activities (ibid.). Indeed, a considerable benefit of UA is the short value-chain involved in having production located so close to where food is consumed.

On the downside, and by virtue of the fact that UA is an activity that frequently operates on the margins of the economy, there are a number of associated risks. First of all, being such densely populated places, cities generate a great deal of liquid and solid waste from both domestic and industrial sources. For farmers on the margins and those requiring inputs to increase their agricultural yields, nutrients are often recycled from waste, leading to significant health risks. For instance, water in Lima polluted with arsenic from upstream mining is used for irrigation, while in Accra and Hyderabad, a long history of waste-water use for agriculture has been chronicled by researchers (ibid.).

A second associated problem is that of raising livestock in heavily urbanized areas. The risk of zoonotic diseases can be high where humans live in close proximity to livestock. This is especially so in some regions of the world where very high peri-urban population density is combined with traditional forms of

Table 15.3
Selected Data on the Contribution of Urban Agriculture to the Availability of Urban Food

City	Weight supplied (per year)	% total supply (city or households)
Nouakchott, Mauritania 1997	11,700 t fruits/vegs	18% vegs/fruits, city
Accra, Ghana 1997	211,000 t vegs (1996) 66,500 t vegs (1992)	90% fresh vegs, city (7.5-66% total food value for producers)
Shanghai, China	1.3 million t vegs (4000 t/d)	60% vegs (100 before) 90% eggs, city
Kumasi, Ghana	150 t/day of fish from farms	13,000 street food kiosks supplied with urban cattle meat, city
Kampala, Uganda		20% of staple food consumption, households 40+% of food for 55% of households 60+% of food for 32% of households 70% of all poultry products consumed in city
Harare, Zimbabwe		60% of food consumption in 25% of poor city households
Bissau, Guinea Bissau		90% leafy vegs, city (grown by 30% households)
Brazzaville, Congo		80% leafy vegs, city (grown by 25% households)

Source: Adapted from Mougeot (2006) and Redwood (2008)

customary animal husbandry. In West Africa, for instance, some cities have experienced incidences of unpasteurized milk use as high as 90 percent, which raises the threat of infection by bovine tuberculosis (Muchaal 2002). To date, no comprehensive epidemiological data have been developed on this topic, but it is clear that the risks of raising livestock in cities should not be underestimated.

A third significant problem is basic access to good land for food production. It is well known that the urban poor have limited land-tenure security for a wide variety of reasons, including the construction of dwellings on vacant public land or private property, construction without permits, or rental without a formal rental contract. The situation of limited land-tenure for the urban poor is exacerbated by the fact that urban land markets occasionally fail due to poor planning, speculation, a lack of access to credit, and unfair competition for access to land (Tannerfeldt and Ljung 2006). Moreover, a new landscape of poverty has emerged in peri-urban areas where agriculture is a common activity, and there are overlapping administrative boundaries between different institutions responsible for the provision of urban services and regulation. The poor are thus pressed to marginal lands—near polluted rivers, waste dumps and flood-prone areas—to pursue their livelihoods. As a livelihood activity, UA can undermine the development of land for dwellings and commercial enterprise. Agriculture, after all, contradicts much of what planners and architects are trained to view as "urban." Urban development no doubt increases the value of land; however, it represents a significant pressure on those dependent on the land for an agricultural livelihood. Planning to protect land for agriculture is difficult to justify to municipalities intent on increasing their tax base or developers wishing to generate rents and profit.

A final weakness worth reporting is rooted in research on UA. Despite dozens of strong case studies and examples of policy action on UA to date, it is still difficult to state the net economic benefit of UA in cities without leaning on dubious data, and there is little consensus. The reason is quite clear: most UA falls outside of formal monitoring or economic data gathering. UA, by its nature and the fact that it is often illegal, is an informal activity. The generation of comparative and robust data using a transferable methodological framework is the next frontier for researchers.

As a result of these problems, many cities are still not fully engaged in supporting UA from a policy standpoint. As the impact of commodity price changes becomes clearer in the coming years, one predicts that UA will become a more popular method to mitigate urban poverty.

The good news is that governments are starting to pay heed to the growing data on UA. Declarations have been signed by senior government ministers (Harare Declaration on UA, 2003) and municipal mayors (Quito Declaration, 2000, signed by forty mayors), illustrating the increased buy-in of decision makers for UA—and such declarations are leading to action. Since 2000, for example, Peru, Brazil, Ghana, and China have created national sectoral policy programs on UA (Redwood 2008). All of the issues outlined above feed into the type of policy used by municipal authorities regarding UA. For instance, in the case of poverty reduction, cities need to explore specific interventions targeting the poor. To increase food availability, a policy that targets wealthier farmers in larger peri-urban farms may be more important. While UA may still be seen primarily as a localized response to changing global food security, the FAO, World Bank, and others are beginning to consider how it may be incorporated into wider national and international policy debates on food security.

The Future of Urban Agriculture

For the time being, food prices have eased, reflecting perhaps the maxim that "there is no better cure for high prices than high prices." It is likely that these changes are temporary since the main drivers outlined above are still pertinent: There is now, however, more time to cope with some of these changes. This is where UA can play a role in promoting food security.

The relative importance of UA is on the rise, both as an economic activity and also as a contribution to environmental management. There arc challenges and gaps that remain, however. First of all, it is incumbent on the research community to pursue better comparative economic data on the value of the UA economy. So far, no comprehensive assessment of the economic value of UA has been undertaken, beyond anecdotal or ad hoc studies (see Resource Centres on Urban Agriculture and Food Security 2003 for a summary of these). This would require surmounting the challenge of data collection and management of what is a largely informal practice. Moreover, quantifying how UA contributes to protecting the poor in this period of volatility is a basic requirement for any proactive evidence-based policy. IDRC is working with the FAO and World Bank on a package of methodologies that could be applied to improve economic data on UA.

Second, UA comes with certain risks and these risks need to be contained If UA is to be proactively developed by city governments. First and foremost is the health risk associated with unsanitary food production in and around cities. The 2006 World Health Organization guidelines on safe irrigation with wastewater provide crucial information on how polluted water is used and can be managed where few clean alternatives exist (WHO 2006).

A third point relates to the often small-scale and informal nature of UA. Despite the fact that there are numerous examples of success in including small-scale farmers in markets, it is difficult to find replicable models and methodologies. The policy challenge is thus how to make successful examples of UA scalable.

Finally, what is necessary is a continued push to legitimize UA where it is still illegal in the eyes of policy-makers and planners and to draw on the positive experiences of some cities to ensure the viability of the UA economy.

Works Cited

Akinbamijo, O., S. Fall, and O. Smith (eds.) (2002). "Advances in Crop-Livestock Integration in West African Cities." Ottawa: International Development Research Centre.

Ba, Diadie (2008). "Hundreds Protest against Food Prices in Senegal." Reuters. 26 April.

Dawe, David (2008). "Have Recent Increases in International Cereal Prices Been Transmitted to Domestic Economies? The Experience in Seven Large Asian countries." Agricultural Development Economics Division, ESA Working Paper No. 08-03. Rome: FAO.

Dixon, Jane, Abiud Omwega, Sharon Friel, Cate Burns, Kelly Donati, and Rachel Carlisle (2007). "The Health Equity Dimensions of Urban Food Systems." *Journal of Urban Health* 84, no. 1: 118–29.

Floquet, Anne, Roch Mongo, and Juste Nansi (2005). "Multiple Functions of Agriculture in Bohicon and Abomey, Benin." *Urban Agriculture Magazine* 15: 9–10.

Food and Agriculture Organization of the United Nations (2002). "The State of Food Insecurity in the World 2001." Rome. http://www.fao.org/docrep/003/Y1500E/Y1500E00.HTM

——— (2008). "Soaring Food Prices: Facts, Perspectives, Impacts and Actions Required." Background paper for the High-Level Conference on World Food Security: the Challenges of Climate Change and Bioenergy. Rome. 3–5 June.

Frayne, Bruce (2005). "Survival of the Poorest: Migration and Food Security in Namibia." In *Agropolis: The Social, Political And Environmental Dimensions of Urban Agriculture*, ed. L.J.A. Mougeot, 31–50. London: Earthscan.

International Development Research Centre (2004) "Optimizing the Use of Vacant Land for Urban Agriculture through Participatory Planning Processes." Final Technical Report, Project number 100983. Ottawa.

——— (2008). "The World Food & Agriculture Crisis." Programs and Policy Brief no. 38. Ottawa.

Kliger, Rachelle (2008). "Cairo Grappling with Bread Crisis." Y.Net News. 6 March.

Koc, Mustafa, Rod McRae, Luc Mougeot, and Jennifer Welsh (1999). *For Hunger Proof Cities*. Ottawa: IDRC.

Mougeot, Luc (2000). "Urban Agriculture: Definition, Presence, Potential and Risks." In *Growing Cities, Growing Food: Urban Agriculture on the Policy Agenda: A Reader on Urban Agriculture*, ed. N. Bakker, M. Dubbeling, S. Gündel, U. Sabel-Koschella, and H. de Zeeuw, 1–42. Feldafing: German Foundation for International Development.

Mubarik, Ali, Hubert de Bon, and Paule Moustier (2005). "Promoting the Multifunctionality of Urban and Periurban Agriculture in Hanoi." *Urban Agriculture Magazine* 15: 11–13.

Muchaal, Pia (2002). "Urban Agriculture and Zoonoses in West Africa: An Assessment of the Potential Impact on Public Health, Cities Feeding People." Report no. 35. Ottawa: International Development Research Centre.

Otte, Joachin, David Roland-Holst, Dirk Pfeiffer, Ricardo Soares-Magalhaes, Jonathan Rushton, Jay Graham, and Ellen Silbergeld (2007). "Industrial Livestock Production and Global Health Risks." Pro-Poor Livestock Policy Initiative Research Report. Rome: FAO.

Oxfam (2008). "Double-Edged Prices." Oxfam Briefing Paper 212. Oxford: Oxfam International.

Redwood, Mark (2008). "Introduction" in *Agriculture and Urban Planning: Generating Livelihoods and Food Security*, ed. M. Redwood. London: Earthscan.

·Resource Centres on Urban Agriculture and Food Security (2003). "Annotated Bibliography on Urban and Periurban Agriculture." Leusden, Sweden. March. http://www.ruaf.org/sites/default/files/annotated_bibliography.pdf.

Santanderau, Alain, and Cecelia Castro (2007). "Social Organisations of Agricultural Producers in Latin America and Europe: Lessons Learned and Challenges." *Urban Agriculture Magazine* 17: 5–7.

Tannerfeldt, Goran, and Per Ljung (2006). *More Urban, Less Poor*. Stockholm: Swedish International Development Cooperation Agency.

World Health Organization (2006). *Safe Use of Wastewater, Excreta and Greywater*, vol. 1–4. Geneva: WHO.

CHAPTER 16

Reorienting Local and Global Food Systems

Institutional Challenges and Policy Options from the UN Agricultural Assessment

Marcia Ishii-Eiteman

We have not so much a "food crisis" today as a "food *system* crisis." The near-collapse of today's local and global food systems has been decades in the making. Over the past several decades, a series of policy interventions (such as structural adjustment policies that dramatically decreased public-sector investments in agriculture and trade liberalization that removed crucial institutional supports for small-scale farmers) along with severe ecological damage to natural and agricultural systems (a by-product of the "green revolution" and shift toward industrialized agriculture over the past fifty years) have compromised the resilience of local food systems and heightened the vulnerability of the rural and urban poor. In countries around the world, local food systems have been increasingly replaced by a globalized food system, governed without much transparency by a few governments and institutional arrangements, and strongly influenced by the small number of transnational corporations that dominate the agricultural inputs, commodity trade, and food retail industries. Yet this global food system has itself proven to be highly vulnerable to social and environmental shocks.

Today's food system crisis is closely linked to today's climate, water, energy, and financial crises. These urgently require new approaches that recognize the root causes and consequences of food price volatility and that aim to enhance social and ecological resilience, reduce vulnerabilities, and increase food sovereignty and environmental sustainability. Accomplishing this requires rethinking our food, agriculture, health, environment, education, and trade policies and practices. Looking ahead, it implies far-reaching structural changes to the current globalized food system and its associated institutions, i.e., the rules, norms, and procedures that guide societal interactions, both formal and informal.

The United Nations and other institutions have recently contributed two important reports that provide guideposts and policy options for the way forward. The 2008 UN Human Rights Council Report (HRCR) of the Special Rapporteur on the Right to Food presents a human rights framework that can be usefully applied to the design and prioritization of policies affecting food security and the right to food, with particular attention to the obligations of states (see de Schutter 2008). The UN-led International Assessment of Agricultural Knowledge, Science and Technology for Development (IAASTD) provides a comprehensive analysis of how existing and emerging agricultural systems, policies, investments, and institutional arrangements can help or hinder efforts to reduce hunger and poverty, improve rural livelihoods, and facilitate equitable and environmentally, socially, and economically sustainable development[1] (see Box 16.1).

Together, these two reports demonstrate that a range of viable policy options exist. New institutions and arrangements will be required. Local engagement in the policy-making process and in the establishment of viable localized and regionalized food systems will be essential. In this context, a rights-based framework would help focus institutional priorities and establish necessary mechanisms to protect rights, in particular those affecting the establishment of equitable

---

Box 16.1   *The UN Agricultural Assessment*

Co-sponsored by the UN Environment and Development Programmes, the Food and Agriculture Organization of the United Nations (FAO), the World Bank, and other institutions,[2] the IAASTD represents two years of rigorous research and synthesis by over four hundred scientists and development experts from over eighty countries, drawn from all regions and multiple disciplines, including biological, physical, and social scientists, public- and private-sector actors, and civil society representatives. Their findings underwent two rigorous external reviews, in which a similarly broad range of experts—over one thousand in total—participated.

After the conclusion of the final intergovernmental plenary in Johannesburg, South Africa, in April 2008, 95 percent of participating governments formally approved the IAASTD report.[3] In a historic moment on the last day of the plenary, all participating governments agreed that the IAASTD represents "a constructive initiative and important contribution that all governments need to take forward" (IAASTD 2009d).

and sustainable food systems. This chapter identifies institutional challenges in reorienting local and global food systems toward equity and sustainability food systems and discusses promising policy options from the two reports.

Role of Institutions[4]

Institutions are the rules of the game that, either formally or informally, make up the policies, procedures, laws, and agreements that shape society. In terms of food systems, relevant institutions include chemical regulations, seed registration and ownership laws, trade rules, corporate "personhood," intellectual property rights, and knowledge rights, as well as gender relations, land tenure rules, community approaches to saving and exchanging seed, water use agreements, and cultural rights. Institutional arrangements represent the formations of actors that agree to implement these institutions. In doing so, they inevitably make choices that reflect their norms, values, and interests. In seeking solutions to the food system crisis, it is therefore particularly important to recognize that while science and technological experience may well inform the choices made by institutional arrangements, they play a largely secondary role to the significant influence of institutional traditions, trajectories, and deep-rooted pre-analytic values and preferences. It is equally important to under stand the historical imbalances in power and influence among these institutional arrangements, imbalances that strongly privilege certain institutional narratives over others (Dreyfus et al. 2009).

Understanding the role of institutions and institutional arrangements in shaping food systems is essential to the identification of new institutional pathways and governance structures that can help reorient food systems to meet sustainable and equitable development goals. An immense challenge lies in reshaping or, where necessary, creating new institutions and arrangements that can more effectively rebalance power and meet diverse societal needs, prioritizing those of vulnerable groups and ecosystems. All this must happen at a time in which society is facing extreme new pressures from climate change, diminishing fossil fuel supplies, water scarcity, and an escalating loss of biodiversity, indigenous peoples, and traditional knowledge systems.

Rights-based Approach

The right to food is a fundamental human right enshrined in the Universal Declaration of Human Rights and several subsequent international treaties and sets of guidelines. As the Special Rapporteur on the Right to Food outlines, the central question for sustainable food systems in human rights terms is: "Who will produce food, how, and for whose benefit?" (de Schutter 2008). This question provides the central unifying thread running through the tapestry of options facing policy-makers; without a clear-sighted response, the possibility of attaining equitable and sustainable food systems unravels.

A human-rights-based approach to the food crisis necessarily focuses on the rights of vulnerable groups to food, water, livelihood, and a safe working environment. Vulnerable groups include those with insecure land tenure, landless labourers, women, displaced and indigenous people, minorities, disabled people, and the rural and urban poor. The right to livelihood and to a safe working environment implies, for farmers, the right to productive resources (e.g., land, water, seed, etc.) and the right not to be poisoned by, for example, toxic chemicals. The right to seed necessarily implies rural and indigenous peoples' right to save and cultivate seed, a practice well established many millennia ago.

With this understanding, the HRCR warns that intellectual property rights (IPR) regimes that prevent farmers from reusing and exchanging seeds can seriously undermine biodiversity and farmers' ability to continue farming, and thus the world's capacity to ensure sustainable food production over the long term. Thus, today's IPR regimes will have to be re-examined and reformulated to avoid threats they may pose to the fulfillment of the right to food. The HRCR also warns policy-makers specifically about focusing too much attention on just increasing production—an over-simplistic, ahistorical approach that mostly benefits transnational agribusinesses. Rather, efforts to address the food crisis must be informed by an understanding of the political economy and political ecology of food, and on how to empower vulnerable groups to realize their rights to food, land, seed, health, and livelihoods.

A human-rights-based framework approach can help focus institutional priorities and establish necessary mechanisms to protect human rights. Adopting such a framework can help state and non-state actors keep food system solutions firmly on track. A framework approach would also help policy-makers avoid getting lost in a laundry list of policy options or distracted by misdirected efforts that may undermine fulfillment of these human rights. This approach can provide a useful frame with which to view the solutions and policy options put forward by the IAASTD. Because of the IAASTD's explicit focus on equity, its analyses and many of its policy options fit well within a human-rights-based framework for action.

IAASTD Key Finding: "Business as Usual is Not an Option"

The IAASTD concludes that today's intertwined crises in climate, energy, water, and food demand change now. While agricultural technological innovations in the twentieth century have contributed to impressive yield gains that have in turn contributed to national efforts to attain food security, their success was largely shaped by immense investments in agriculture and an array of institutional and policy supports (Dreyfus et al. 2009). At the same time, the technological contributions of the Green Revolution and industrialized agriculture have had unacceptably high social and environmental costs, including natural resource degradation, salinization, and desertification, rising water scarcity, and chemical pollution leading to dead zones at sea, groundwater contamination, loss of biodiversity, and public health harms.

People have also benefited unevenly from yield gains of the past decades. As the IAASTD explains, "agricultural technologies such as high-yielding crop varieties, agrochemicals, etc., have primarily benefited better resourced groups in society and transnational corporations, rather than the most vulnerable ones" (IAASTD 2009d, 32). As a result, social inequities have deepened over the past several decades, and women, children, the rural and urban poor, farm workers, and indigenous communities have suffered disproportionately. The report explains that reliance on technological solutions—including genetically modified seeds—is unlikely to reduce hunger or poverty or to advance equitable and sustainable development, and may—in the absence of careful comprehensive and participatory assessment of technological innovations by multiple stakeholders—even exacerbate current conditions of poverty and inequity. The more fundamental change required must come from new political, social, economic, and cultural approaches to the institutions governing not only those technologies but also the evolution of our food and agricultural systems, IPR frameworks, and international trade rules (McIntyre et al. 2009a).

In assessing existing institutional arrangements, the IAASTD finds that corporate concentration within the food and agribusiness industry, and vertical integration of the food system, have had negative consequences for health, environment, and social equity (McIntyre et al. 2009b; Dreyfus et al. 2009). The evidence demonstrates, furthermore, that trade liberalization has more often than not harmed the poorest countries, peoples, and their environments and poses a serious threat to food security (Nathan, Rosenthal, and Kagwanja 2009; Izac et al. 2009) (see Box 16.2).

IAASTD Key Policy Directions: Reorient toward Sustainability

The IAASTD lays out a comprehensive set of social, environmental, and economic policy options to reorient local and global food systems toward greater social equity and sustainability. In brief, these include: strengthening the small-scale farm sector; revitalizing local and regional food systems; building local

> **Box 16.2** *IAASTD Key Finding on Trade*
>
> "Opening national agricultural markets to international competition can offer economic benefits, but can lead to long term negative effects on poverty alleviation, food security and the environment, without basic national institutions and infrastructure being in place. Some developing countries with large export sectors have achieved aggregate gains in GDP, although their small-scale farm sectors have not necessarily benefited and in many cases have lost out. The small-scale farm sector in the poorest developing countries is a net loser under most trade liberalization scenarios" (IAASTD 2009d, 7).

and national capacity in biodiverse, agroecological farming; mobilizing public and private-sector investments toward equitable sustainable development; and establishing supportive institutions and institutional arrangements to accomplish these goals. The IAASTD underscored the importance of recognizing the right of peoples and countries to determine their own food and agricultural policies, defined as food sovereignty (see also Ishii-Eiteman 2009; Via Campesina 1996, 2007; IIED 2008).

Social Policy Options

Support small-scale farmers

The IAASTD states that establishing equitable and sustainable development in the future requires prioritizing the needs of small-scale farmers now (Izac et al. 2009; Dreyfus et al. 2009). Institutional and policy options to accomplish this include increasing public investments in rural areas and strengthening farmers', women's, and other community-based organizations; providing technical assistance to farmers in adjusting to and mitigating climate change and other environmental stresses and system shocks; and encouraging equitable and participatory farmer–scientist partnerships to respond more appropriately to farmers' immediate and emerging challenges. Small-scale farmers also need secure access to productive resources (e.g., land, water, seeds), information, and credit and marketing infrastructure as well as fair trade arrangements and supportive market conditions. The IAASTD observes that intellectual property laws will also likely need to be revised to prevent the misappropriation of indigenous and local people's knowledge and to more effectively address equity and genetic resource issues. This revision will also need to address tensions between, on the one hand, traditional knowledge, rights, and community-based innovation, and corporate ownership claims on DNA, germplasm, seeds, and other biological components or forms of life on the other (Izac et al. 2009).

Revitalize local and regional food systems

Strengthening local and regional food systems offers a compelling pathway toward achieving equitable and energy-efficient food production and distribution. This can be stimulated through innovative multi-stakeholder partnerships between rural and urban communities, private-sector representatives, and public-sector agencies. Promising approaches include the establishment of representative democratic local and state food-policy councils as a means of encouraging broad participation in setting food policies (as in Canada, India, the Netherlands, the United Kingdom, and the United States) and the encouragement of urban and peri-urban agriculture, an increasingly important component of food security in many countries (for instance, in municipalities in Brazil, China, Cuba, Kenya, India, Uganda, Venezuela, and Vietnam).

Localization or regionalization of food processing, procurement (for example by schools, hospitals, city agencies, etc.), and distribution are examples of innovative approaches. The municipality of Belo Horizonte in Brazil has implemented a UN-award-winning food security program, procuring mostly organic produce from nearby small-scale farmers and supplying urban needs, while farm-to-school and other private-sector-initiated local procurement programs are gaining popularity in North America. Similarly, relying as much as possible on local or regional resources for emergency food distribution systems as an alternative to internationally sourced food aid can both reduce energy costs and support local and regional agricultural sectors.

Environmental Policy Options

Recognize the multi-functionality of agriculture

The IAASTD has determined that agriculture is multi-functional in nature, providing goods and services that reflect the interconnectedness of agriculture's multiple dimensions, roles, and functions (IAASTD 2009d, 5; McIntyre et al. 2009b; Leakey et al. 2009; IAASTD 2009f) (see Box 16.3). Thus, institutions and institutional arrangements need to be closely assessed for their contributions and potential impacts, both positive and negative, to the multiple functions of agriculture. Public and private-sector actors may need to reformulate their programs and policies to ensure that the multiple functions of agriculture are sustained, and that public interest and natural and agricultural resource management goals are met.

The IAASTD goes on to elaborate the types of policies and investments that can secure the social, environmental, and economic functions of agriculture. Social functionality can be ensured by empowering marginalized stakeholders (particularly women) to sustain the cultural and biological diversity of their food and agriculture systems and by increasing their access, ownership,

---

Box 16.3 *Multi-functionality of Agriculture*

According to the IAASTD:

"Agriculture operates within complex systems and is multi-functional in its nature. A multi-functional approach to implementing agricultural knowledge, science, and technology (AKST) will enhance its impact on hunger and poverty, improving human nutrition and livelihoods in an equitable, environmentally, socially and economically sustainable manner. Multi-functionality recognizes the inescapable interconnectedness of agriculture's different roles and functions, i.e., agriculture is a multi-output activity producing not only commodities, but also non-commodity outputs such as environmental services, landscape amenities, and cultural heritages" (IAASTD 2009f; see also IAASTD 2009d).

---

and control of economic and natural resources through legal and financial (credit) means. Education and training—not only of farmers but also of policy-makers and public-agency personnel—is necessary to strengthen decentralized participatory planning and decision-making processes. Environmental functionality is supported through the adoption of biodiverse, agroecologically sound practices, restoration and protection of ecosystem services (water and nutrient cycles, pollinator health, etc.), and use of water and energy-conserving practices. Adverse effects of climate change can be minimized and mitigated through diversified farming practices, such as organic agriculture, which support increased soil carbon sequestration and water retention. Economic functionality can be secured by promoting market and trade policies that benefit small-scale producers, establishing fair price bands and increasing access to microcredit, rewarding resource-conserving practices, and providing access to crop insurance and other financial services and instruments.

An integrated approach to the multi-functionality of agriculture is likely to yield the best outcomes. This approach calls for including sciences and fields of expertise outside conventional agriculture disciplines (i.e., social sciences, political economy, political ecology, macroeconomics, etc.), bringing multiple ministries together into new institutional formations, and drawing on non-formal science and local and traditional knowledge when devising integrated food, agricultural, and natural resource policies. The IAASTD concludes that broadening agricultural research objectives to address multi-functional goals and revising investment priorities accordingly can substantially improve the multi-functional performance of food and agriculture systems in all parts of the world.

Build capacity in agroecology

The emerging consensus reflected in the IAASTD is that the success of future agriculture will be determined largely by our capacity to adapt to expected and unexpected shocks to the system. Food system impact analyses will thus increasingly need to take account of global water, energy, and climate "foodprints." The central scientific and technical challenge facing agriculture today, according to the IAASTD, is shifting toward improved and sustainable production based on long-term agroecosystem health and ecological resilience in the face of these stresses. The IAASTD therefore calls for "an increase and strengthening of investments in the agroecological sciences" (IAASTD 2009d) and suggests that governments consider establishing a national framework for the implementation of agroecological production (IAASTD 2009d, 6; see also Leakey et al. 2009; Dreyfus et al. 2009; Gurib-Fakim et al. 2009; IAASTD 2009e; PANNA 2009) (see Box 16.4).

Box 16.4 *Agroecology and Sustainable Production*[5]

Agroecology—the foundation of sustainable agriculture—is the science and practice of applying ecological concepts and principles to the study, design, and management of sustainable agroecosystems. Agroecology combines scientific inquiry with indigenous and community-based experimentation, emphasizing technology and innovations that are knowledge-intensive, low-cost and readily adaptable by small- and medium-scale producers. Drawing on both natural and social sciences, agroecology provides a framework for assessing four key systems properties of agriculture: productivity, resilience, sustainability, and equity.

Agroecological farming can increase ecological resilience to environmental shocks such as climate change, improve health and nutrition through decreased exposure to pesticides and improved dietary diversity, increase energy efficiency through reduced reliance on fossil fuels, and conserve resources and essential ecosystem services (e.g., water and nutrient cycling, pollination, natural biological control of pests, maintenance of genetic diversity, prevention of soil erosion, etc.). In Central America, for example, in the aftermath of Hurricane Mitch, farmers who had adopted agroecological methods enjoyed higher productivity, retained more topsoil, field moisture, and vegetation on their plots, and experienced lower economic losses than conventional farmers, demonstrating greater resilience to extreme weather events (Niva et al. 2009).

Specific steps for building local and national capacity in agroecology include increasing investments in agroecological research, extension and education, and encouraging collaboration among farmers, indigenous communities, extensionists, and researchers. Payment incentive programs can encourage practices that increase agro-biodiversity while taxes on health and environmental harms can help reduce reliance on chemical inputs, fossil fuels, and water, or energy-intensive production (Izac et al. 2009; Beintema et al. 2009). Relevant actors will need to revise institutional priorities, incentive systems, and budget allocations to achieve these goals.

The IAASTD also recognizes the importance of minimizing environmental harms caused by agriculture through environmental regulations and the ratification and implementation of regional and international environmental agreements (for example, the Kyoto Protocol to the UN Framework Convention on Climate Change, the UN Convention on Biological Diversity, and the Basel, Stockholm, and Rotterdam Conventions on the registration, trade, use, and reduction of toxic chemicals). Voluntary standards (the Food and Agriculture Organization Code of Conduct on the Distribution and Use of Pesticides), policy frameworks (the Strategic Approach to International Chemicals Management), and other intergovernmental and multi-stakeholder forums to guide policy-making (the International Forum on Chemical Safety) are also effective mechanisms to support national policy transitions toward sustainable development goals. By establishing that reliance on environmentally destructive agricultural practices is no longer acceptable and that continued contribution to greenhouse gas emissions, biodiversity loss, or chemical contamination of biota, air, soil, and water is to be halted, these international agreements can strengthen national political will and commitment to transitioning toward more sustainable practices (Dreyfus et al. 2009; UNESCO/SCOPE/UNEP 2009).

Economic Policy Options

Address food price volatility

Short-term measures to address food-system needs include stabilization of prices (i.e., through fair price floors and ceilings, re-establishment of strategic grain reserves, and supply management mechanisms). By reducing volatility in food prices, these measures can encourage farmers to invest in longer-term resource-conserving strategies, which also support national food security goals. Some short-term responses such as lowering food import tariffs and imposing export restrictions can bring consumers immediate relief but can have longer term costs on domestic food production. Food import tariffs prevent domestic farm sectors from being undermined by food dumping and provide much-needed revenue to public budgets. For this reason, the IAASTD recognizes the need to provide compensation to developing countries for revenues lost as a result of tariff reductions.

The IAASTD also argues for more comprehensive and fully participatory assessment of future investments in biofuels (Avato, Brown, and Kairo 2009; IAASTD 2009a). The HRCR goes further in decrying industrial-scale agrofuels, calling for a suspension of national investments in agrofuels until it can be demonstrated that vulnerable communities and ecosystems will experience no adverse effects.[6] The HRCR also calls for revising current measures that encourage agrofuel production (e.g., blending mandates, subsidies, and tax breaks) (de Schutter 2008).

Implement equitable trade and market-oriented policies

After assessing the adverse impacts of current trade regimes, the IAASTD presents a number of options for how to establish fair regional and global trade arrangements (Nathan et al. 2009; Izac et al. 2009; IAASTD 2009d, 7; and IAASTD 2009c). Foremost among these is preserving national policy flexibility by according special and differential treatment to developing countries. The IAASTD has determined that this will contribute to improving developing countries' ability to benefit from agricultural trade, pursue food security goals, and minimize dislocations from trade liberalization. Providing developing countries with preferential (non-reciprocal) access to industrialized-country markets for commodities important to domestic food and livelihood security, and removing escalating import tariffs for processed commodities, can enable developing countries to gain a fair share of value-added benefits from the export of processed commodities.

These approaches necessitate improving the quality and transparency of governance in agricultural trade, including strengthening developing-country capacities in trade analysis and negotiation. Developing and providing improved tools for assessing social, environmental, and economic tradeoffs in proposed trade agreements (such as strategic impact assessments) is an essential step toward accomplishing this goal.

Market-oriented public-policy options to reorient food systems toward sustainability include the provision of incentives, such as payments for ecosystem services and for organic transitions, along with credit, crop insurance, and tax exemptions for sustainable practices. Public investment in local agro-processing and marketing infrastructure enables value-addition and creates off-farm rural jobs. Public-policy initiatives can facilitate direct farmer-to-consumer sales, for example, by providing infrastructure for urban farmers' markets. Other promising options include encouraging geographic, fair trade, and sustainable production labels, enacting laws that support consumers' right to know about the economic, environmental, and social conditions behind production and distribution, and ensuring availability of affordable third-party certification. This can increase opportunities for commercializing sustainably produced goods. Unsustainable practices can be reduced by levying taxes on health and environmental harms (e.g., the "polluter pays" principle), and carbon and energy taxes based on whole-system energy budgets and analysis of greenhouse gas emissions.

Significantly, the IAASTD also highlights the importance of developing and implementing full-cost accounting measures that include the full array of health, energy, and environmental costs ("externalities") and "spillover" effects associated with different food and agricultural systems (IAASTD 2009b). Obtaining an accurate estimate of the true costs of production is not only good economic practice but essential to enable well-informed policy and budgetary decisions.

Investing in Sustainable Food Systems

Establishing the programs and policies described above would not be without cost. As a first step, policy-makers are encouraged to conduct a participatory multistakeholder assessment of the full costs of food and agricultural systems, programs, and public investments across sectors, calculated over different time scales (IAASTD 2009d, 7; Beintema et al. 2009). This would inform the identification of societal priorities and particular needs of vulnerable groups. A gradual redirection of investments from costly programs with high externalities toward those likely to advance long-term health, environment, and food security goals can minimize system disruptions. Such a rethinking of investment will often benefit from cross-sectoral collaboration, for instance, between ministries and departments of health, agriculture, environment, education, labour, etc. Upfront transaction costs may not be insignificant, but in most cases they will ultimately be offset by reduced externalities and more efficient attainment of policy objectives. International development and donor agencies have a responsibility to assist developing countries with these transitional transaction costs.

Additional public revenues can be generated by taxing health and environmental harms as described above and by levying equitably adjusted user fees, such as for water use and school lunch programs (Nathan et al. 2009). Significant savings can often be secured by removing unnecessary or distorting budget allocations, such as high export production subsidies in industrialized countries and exemptions on import duties and on sales taxes for inputs such as synthetic pesticides. Similarly, in some developing countries, savings can also be accumulated by removing large, anticipatory "pest outbreak" budgets, which become less necessary with the establishment of sustainable pest-management approaches.

The impact of public investments can be substantially strengthened through appropriate mobilization of the private sector. Rewarding private investment in safe, sustainable, and locally appropriate crops, seed systems, technologies, and food markets through tax breaks, for example, can stimulate private-sector engagement.

At the same time, the public sector needs to ensure that impacts of private investments actually benefit the health and food security of all. Public institutional arrangements can initiate competitive bidding for public funding based on an enterprise's proven capacity to meet public-interest goals. They can also establish and enforce codes of conduct to prevent conflict of interest and strengthen corporate accountability both to shareholders and to the public, where public–private partnerships are concerned (IAASTD 2009d, 7–8). Implementation of antitrust and competition regulations can begin to counter some of the adverse effects associated with increasing concentration and vertical integration of the global food system (Izac et al. 2009).

Transnational buyers (trading companies, agrifood processors, input manufacturers) typically dominate globalized food chains. Primary producers capture only a fraction of the international price of a traded commodity. Building countervailing negotiating power, through new institutional arrangements such as farmer co-ops and networks, for example, provide important opportunities for resource-poor farmers to increase their share of "value-added" or "value-captured." The establishment of mechanisms for local rural enterprises to increase their share of value-added (for example through local agro-processing facilities) can also be effective (ibid.).

Institutional Innovations for Improved Governance: Highlights

The IAASTD and HRCR offer a wealth of options for improving the governance of institutions and their associated institutional arrangements to achieve more equitable and sustainable food systems. Some of these have been described above. Four priority areas for institutional improvement, and examples of specific options for action indicated in the IAASTD,[7] are detailed below.

Promote Institutional and Policy Innovations
Reform allocation of ministerial responsibilities
Reforming the allocation of ministerial responsibilities will facilitate more integrated and therefore effective political decisions regarding food and agriculture. Examples of Costa Rica's Ministry of Environment, Energy, Mines, Water, and Natural Resources, and the United Kingdom's Department of Food, Environment,

and Rural Affairs are given in the IAASTD report. In Costa Rica's case, the integration of formerly separate ministerial domains enabled policy-makers to work together to implement a more holistic systems-oriented approach. As a result, 98 percent of Costa Rican energy is now produced from renewable sources and agricultural policies address water scarcity.

Establish a national framework for the implementation of agroecological production

Restructuring agricultural research, extension, and education systems and provide professional and financial incentives will facilitate institutional redirection of resources toward agroecological sciences, integrated natural resource management approaches, and interdisciplinary, farmer-participatory research programs.

Use full-cost accounting methods in comparative assessments of agriculture and food systems

This ensures a more accurate reading of the true costs of food and agricultural industries than is available using most standard economic models. Internalizing the costs of social and environmental "externalities" can also be partly accomplished by levying taxes on harms and by stimulating market response through product labelling (e.g., calculating a product's contribution to greenhouse gas emissions or as a climate footprint).

Employ Comparative Technology Assessments (CTAs)

Comparative Technology Assessments (CTAs) enable governments to monitor and evaluate major new technologies and their socioeconomic, health, and environmental impacts. Institutional arrangements that could be revitalized to coordinate such work at the international level include the UN System to Conduct Technology Assessment for Development and the UN Commission on Science and Technology for Development (within the UN Economic and Social Council). Alternatively, a legally binding multilateral agreement on CTA could be negotiated, providing an independent transparent early warning, monitoring, and assessment framework for emerging technologies and their potential impact on food systems, poverty, equity, environment, and so on.

Reform Commodity Trade and Markets

Establish fairer regional and international trade arrangements

These include arrangements based on recognition of principles of special and differential treatment, non-reciprocal access, and deeper preferential access to developed-country markets for commodities significant to producers' livelihood and food security.

Build developing countries' trade-negotiating capacities

Developing-country negotiators need access to analytical tools and comprehensive information on potential social and environmental tradeoffs associated with trade agreements under consideration. For example, strategic impact assessments of trade agreements give negotiators (and the public) a more comprehensive understanding of potential social and environmental consequences of such agreements.

Establish international competition policy

This policy should create multilateral rules on restrictive business practices and reintroduce price bands to minimize world price volatility (see for example, the Chile–EU trade agreement). Establishing a new UN agency to carry on the important work of the UN Centre for Transnational Corporations could be helpful in this regard.

Revise Laws of Ownership and Access

Implement effective policies for equitable access to resources

Land reform, equitable water use policies, and mass distribution of credit can secure small-scale farmers' access to productive resources. Improved access to productive resources, technologies, equipment, and information can enable small-scale farmers to compete more effectively and invest in longer-term resource-conserving strategies.

Revise intellectual property laws

New laws regarding intellectual property are needed to address the increasing concentration of control over seeds, the declines in agro-biodiversity, and the geographic and social inequities in access to germplasm and knowledge that have resulted.

The new laws should strengthen legal and social protections for indigenous peoples and their knowledge systems, and also uphold small-scale farmers' rights to domesticate, save, develop, exchange, and trade their communities' own agricultural genetic resources. These protections maintain cultural and biological diversity and improve family nutrition, livelihood security, and farm-system resilience to climate change and other environmental stresses.

Broaden Governance and Management of Food and Agricultural Systems

The following are a number of suggested reforms to broaden the governance and management of food and agricultural systems:

- Open up scientific and technological direction-setting processes to a broad array of stakeholders. Such an opening would help ensure that agricultural research efforts are geared toward meeting public interest and environmentally sustainable goals.
- Establish democratic, representative food-policy councils at local, subnational and national levels. These councils ensure that food and agricultural policies and institutions meet broadly agreed-upon societal goals.
- Localize and/or regionalize food processing, procurement, distribution, and consumption arrangements.
- Eliminate structural overproduction in developed countries and the dumping of food below cost of local production in developing countries.
- Redesign emergency food distribution systems to secure goods from local and regional sources.
- Ensure that public- and private-sector partnerships and arrangements serve public interest goals (for example, by providing financial rewards and tax breaks for good practices, taxing harmful practices, and establishing and enforcing anti-trust and international competition regulations, along with codes of conduct and competitive bidding where public-sector resources are involved).

Rights-based Approach to Food Security: Reprise

In answering the question posed earlier, "Who will produce food, how, and for whose benefit?" (de Schutter 2008), the rights-based approach offers several institutional mechanisms to ensure that the policy options suggested by the IAASTD have a high likelihood of advancing the more equitable development of our food systems. They begin by identifying emerging or potential threats to the right to food by assessing, for example, the impacts of legislative initiatives, IPR regimes, trade agreements, and other policies with direct or indirect effects on food security. The HRCR suggests "right-to-food impact assessments" and notes that food sovereignty is a precondition of food security. The HRCR also addresses the thorny question of accountability and how to provide redress for legal claims against those whose action—or inaction—violates these rights. The report notes that institutional mechanisms for accountability may require new legislation to ensure that such rights are justiciable—actionable in court, with redress and liability enforcement mechanisms in place.

Furthermore, as the HRCR reminds us, it is the obligation—of states and all international agencies—*not* to pursue policies that could negatively affect fulfillment of the right to food, such as the development of agrofuel policies. These actors and institutional arrangements are also bound by the obligation to protect rights "by controlling private actors" and should therefore clarify how the private sector can contribute to shaping a more just food production and distribution system. Finally, all states, international agencies, and their members must cooperate internationally to address not only the proximate but also the underlying structural causes of the food crisis and failures of our food systems (ibid.).

Where the IAASTD has provided us with a detailed set of robust policy options, the HRCR offers a rights-based framework that bolsters these policy options with the imperative of fulfilling the universally recognized right to food.

Conclusion

By strengthening farmers' organizations, supporting the small-scale farm sector, increasing investments in agroecological farming, creating more equitable and transparent trade agreements, and increasing local participation in policy formation and other decision-making processes, we can begin to reverse the structural inequities within and between countries, increase rural communities' access to and control over resources, and pave the way toward local and national food sovereignty. Many examples of successful approaches already exist in the world but are often inadequately supported by prevailing national and international policy and trade environments. Implementing the IAASTD and HCRC's robust options for the future requires governments, international agencies, and the United Nations to recognize their obligations to respect the human right to food and to take decisive action in setting a new course for food and agriculture that fulfills the promise of equitable and sustainable development.

Notes

1   The full IAASTD report includes a Global Report and five sub-global reports and their respective Summaries for Decision Makers as well as a Synthesis Report, including an Executive Summary. The reports were accepted at an Intergovernmental Plenary in Johannesburg in April 2008.
2   The assessment was sponsored by the United Nations, the World Bank, and the Global Environment Facility (GEF). Five UN agencies participated: the Food and Agriculture Organization of the United Nations (FAO), the UN Development Programme (UNDP), the UN Environment Programme (UNEP), the UN Education, Scientific and Cultural Organization (UNESCO) and the World Health Organization (WHO).
3   Fifty-eight governments accepted and endorsed the report. Only the United States, Canada, and Australia declined to fully approve the report, objecting primarily to IAASTD findings on the impacts of trade liberalization and its assessment of modern biotechnology's actual and potential contribution to equitable and sustainable development goals.
4   For further background, see IAASTD 2009b.
5   For further background, see IAASTD 2009f.

6 The HRCR, along with many civil society organizations and social movements, use the term "agrofuels" to emphasize the industrial scale and purpose of both the production system and its outputs. The term "biofuels" is considered misleading as, some argue, it incorrectly implies a "green" or environmentally sustainable form of energy production.
7 For details on the listed options for institutional change, see in particular, McIntyre et al. 2009b; Izac et al. 2009; Nathan et al. 2009; IAASTD 2009b; and IAASTD 2009c.

Works Cited

Avato, P., R. Brown, and M. Kairo (2009). "Bioenergy." In McIntyre et al. 2009b, 35–39.

Beintema, Neinke, A. Koc, P. Anandajayasekeram, A. Isinika, F. Kimmins, W. Negatu, D. Osgood, C. Pray, M. Rivera-Ferre, V. Santhakumar, H. Waibel, J. Anderson, S. Dehmer, V. Gottret, P. Heisey, and P. Pardey (2009). "Agricultural Knowledge, Science and Technology: Investment and Economic Returns." In McIntyre et al. 2009a, 495–550.

De Schutter, Olivier (2008). "Building Resilience: A Human Rights Framework for Food and Nutritional Security: Report of the Special Rapporteur on the Right to Food." A/HRC/9/23. Geneva: United Nations High Commissioner for Refugees. September.

Dreyfus, F., C. Plecovich, M. Petit, H. Acka, S. Dogheim, M. Ishii-Eiteman, R. Kingamkono, J.L.S. Jiggins, and D. Keith (2009). "Historical Analysis of the Effectiveness of AKST Systems in Promoting Innovation." In McIntyre et al. 2009a, 57–144.

Gurib-Fakim, A., L. Smith, N. Acikgoz, P. Avato, D. Bossio, K. Ebi, A. Gonçalves, J. Heinemann, T. Martina Herrmann, J. Padgham, J. Pennarz, U. Scheidegger, L. Sebastian, M. Taboada, E. Viglizzo, F. Bachmann, B. Best, J. Brossier, C. Farnworth, C. Gewa, E. Gyasi, C. Izaurralde, R. Leakey, J. Long, S. McGuire, P. Meier, I. Perfecto, and C. Zundel (2009). "Options to Enhance the Impact of AKST on Development and Sustainability Goals." In McIntyre et al. 2009a, 377–440.

International Assessment of Agricultural Knowledge, Science and Technology for Development (2009a). "Bioenergy and Biofuels: Opportunities and Constraints." Washington, DC. http://agassessment.org/docs/10505_Bioenergy.pdf.

——— (2009b). "Business as Usual is Not an Option: The Role of Institutions." Washington, DC. http://iaastd.net/docs/10505_Institutions.pdf.

——— (2009c). "Business as Usual is Not an Option: Trade and Markets." Washington, DC. http://iaastd.net/docs/10505_Trade.pdf.

——— (2009d). "Summary for Decision Makers of the Global Report." Washington, DC: Island Press.

——— (2009e). "Summary for Decision Makers of the Latin America and the Caribbean (LAC) Report." Washington, DC: Island Press.

——— (2009f). "Towards Multifunctional Agriculture for Social, Environmental and Economic Sustainability." Washington, DC. http://iaastd.net/docs/10505_Multi.pdf.

International Institute for Environment and Development (2008). "Towards Food Sovereignty: Democratising the Governance of Food Systems." London. http://www.iied.org/natural-resources/key-issues/food-and-agriculture/towards-food-sovereignty-democratising-governance-food-systems#resources.

Ishii-Eiteman, Marcia (2009). "Food Sovereignty and the International Assessment of Agricultural Knowledge, Science and Technology for Development." *Journal of Peasant Studies* 36, no. 3.

Izac, Anne-Marie, H. Egelyng, G. Ferreira, E. Acheampong, S. Barabosa, D. Duthie, D. Hautea, B. Hubert, N. Louwaars, M. McLean, S. Suppan, M. Wierup, and M. Wilson (2009). "Options for Enabling Policies and Regulatory Environments." In McIntyre et al. 2009a, 441–94.

Leakey, Roger, G. Kranjac-Berisavljevic, W. Abedini, S. Afiff, M. Ahmed, M. Ali, N. Bakurin, S. Bass, P. Caron, P. Craufurd, A. Hilbeck, T. Jansen, S. Lhaloui, K. Lock, A. Martin, A.J. McDonald, J. Newman, O. Primavesi, T. Sengooba, A. Lucie Wack, and S. Suppan (2009). "Impacts of AKST on Development and Sustainability Goals." In McIntyre et al. 2009a, 145–254.

McIntyre, Beverly D., Hans R. Herren, Judi Wakhungu, and Robert T. Watson, eds. (2009a). *International Assessment of Agricultural Knowledge, Science and Technology for Development: Global Report.* Washington, DC: Island Press.

———— (2009b). *International Assessment of Agricultural Knowledge, Science and Technology for Development: Synthesis Report.* Washington, DC: Island Press.

Nathan, Dev, E. Rosenthal, J. Kagwanja (2009). "Trade and Markets." In McIntyre et al. 2009b.

Niva, Elsia, I. Perfecto, M. Ahumada, K. Luz, R. Pérez, and J. Santamaría (2009). "Agriculture in Latin America and the Caribbean: Context, Evolution and Current Situation." In McIntyre et al. 2009b, 1–74.

Pesticide Action Network North America (2009). "Agroecology and Sustainable Development: Findings from the UN-led International Assessment of Agricultural Knowledge, Science and Technology for Development." San Francisco. http://www.panna.org/jt.

United Nations Educational, Scientific and Cultural Organization, Scientific Committee on Problems of the Environment, United Nations Environment Programme (2009). "Towards Sustainable Agriculture: Some Key Findings of the International Assessment of Agricultural Knowledge, Science and Technology for Development." UNESCO-SCOPE-UNEP Policy Brief Series, No. 8. Paris. March.

Via Campesina (1996). "The Right to Produce and Access to Land." Rome. November 11-17. http://www.voiceoftheturtle.org/library/1996%20Declaration%20of%20Food%20Sovereignty.pdf.

———— (2007). "Nyéléni Declaration." Sélingué, Mali: World Forum on Food Sovereignty.

CHAPTER 17

The Governance Challenges of Improving Global Food Security

Alex McCalla

F ood security is a national responsibility. The definition of the Food and Agriculture Organization of the United Nations (FAO) states: "Food security exists when all people, at all times, have physical, social and economic access to sufficient, safe, and nutritious food which meets their dietary needs and food preferences for an active and healthy life" (FAO, 2003). Therefore, the sufficient condition for food security for an individual is accepted as having three necessary conditions: (1) availability of sufficient supplies; (2) access to enough food, either through sufficient income or by the provision of adequate safety nets; and (3) nutritional wholeness—access to a complete and healthy diet. The sufficient conditions for national food security are that all citizens are individually food secure.

Historically, many, including most agriculturalists, have equated food security to national food self-sufficiency but this is clearly inadequate. Even if a nation were to produce on average enough food for all of its citizens, there is no guarantee that all citizens have access to it or that they would consume a nutritious diet. But beyond this, the reality is that no country in the world is truly food self-sufficient. The United States, one of the world's largest exporters of agricultural products, is forecast to export US$113 billion in 2009, but is also forecast to purchase $US83 billion of agricultural imports in the same year. By comparison, Japan, which, based on calories consumed, imports 60 percent of its food supply, imported $US40 billion of food in 2007. Therefore every country is involved to some extent in international food markets, and international governance must play a role.

This chapter focuses on four types of international governance issues. First, feeding a growing world population, given limited land and water resources, has increasingly depended on increased yields. Increased yields come from agricultural productivity improvements generated by agricultural research and development (R&D). One of the most successful innovations in international governance is the Consultative Group on International Agricultural Research (CGIAR), which played a critical role in the "green revolution." Knowledge does not recognize political boundaries, and, therefore, international mechanisms are crucial. Yet the CGIAR faces serious challenges from both changes in science and changes in the structure of international aid.

Second, all countries are involved to some degree in international trade in agricultural and food products. Therefore an efficient, open, and fair trading system is clearly an international "public good" of critical importance to all countries. Trade liberalization under the General Agreement on Tariffs and Trade (GATT) and its successor organization, the World Trade Organization (WTO), has been very successful in liberalizing industrial trade but has had limited success in agriculture. Therefore agricultural trade liberalization remains a stubborn international governance challenge.

Third, the more important international markets are to domestic food security, the more critical it is for countries to manage their interface with world markets. This prevents wide swings in international prices from damaging domestic rural development and safety net programs. This does not mean closing borders but rather developing national, and possibly regional, coping mechanisms that could involve stocks and other risk-management tools.

Fourth, there will always be some at the national and global levels who have inadequate access to food. Therefore national and international safety nets, like food distribution programs, food stamps, and income supplements, are critical elements in long-term food security strategies. International food aid, as through the World Food Programme (WFP), for example, has a critical role to play in regional food shortages resulting from natural disasters such as drought and national food crises resulting from conflict or other contributing factors. After considering each of these issues in turn, the chapter concludes by assessing the overall prospects of improving international governance for food security.

The Critical Importance of Global Agricultural Productivity Improvement

Past trajectories regarding human population growth and food production make it clear that improvements to agricultural productivity are required and that this must be a central focus of international governance responses to the food crisis.

In 1961, the world's population reached 3 billion people, increasing from 2 to 3 billion in just thirty-four years. In contrast, the second billion had taken 123 years (1804–1927) and the first billion from the beginning of time. According to Evans, virtually all of the increased food production needed to feed the first 2 billion came from expanded area under production. And despite pockets of scientific agriculture in Western Europe and Japan in the nineteenth century, the third billion was likewise primarily fed by a 40 percent increase in area and from the freeing of 130 million hectares previously producing fuel for horses and for food grain production. "Between 1870 and 1920, while world population increased by 40 percent, the arable area increased by 75 percent due to extensive land clearing, particularly in North America and Russia" (Evans 1998, 90).

It was only after 1960 that increasing yields per hectare became the major source of increases in food supplies. Adding the fourth and fifth billion took only thirteen years each (1961–74 and 1974–87), and the sixth just twelve years (1987–99). The vast majority of the increase in food production needed to feed this doubling of world population in less than 40 years came from increased productivity, as modest increases in area since 1975 were more than offset by losses of productive land to other uses and to soil degradation. The application of science to agriculture had research roots dating back at least to von Liebig in the mid-nineteenth century, but it was increasing investments in applied research in developed countries in the first half of the twentieth century that led to the genetic and chemical revolution that drove agriculture in the second half of the twentieth century. Molecular biology will shape improvements in the twenty-first century. No matter your position on it, this is an incredible scientific accomplishment.

But the task is not complete. The world's population is forecast to be 8.9 billion people in 2050, which is 2.2 billion more than the current population of 6.7 billion. Rural population numbers will stabilize and likely decline, meaning that all of the increase will be in cities. These additional people will have to be fed using less land and with more severe competition for water. The recent price spike should remind us how precarious the balance between the rates of growth of supply and demand has become. Global grain consumption exceeded global production in six of the last eight years (Figure 17.1). The result was a drawdown of stocks to the lowest levels since the early 1970s (Figure 17.2). Thus, when shocks like weather and the surge in biofuel demand occurred, they caused prices to rise sharply.

The major reason the rate of supply expansion is slowing is because productivity growth in basic cereals is declining, as shown in Figure 17.3. In the early 1960s, productivity growth for the basic cereals, maize, rice, and wheat was 3 percent or above (the beginnings of the green revolution). In the 1970s

Figure 17.1
World Cereal Production, 1999–2007

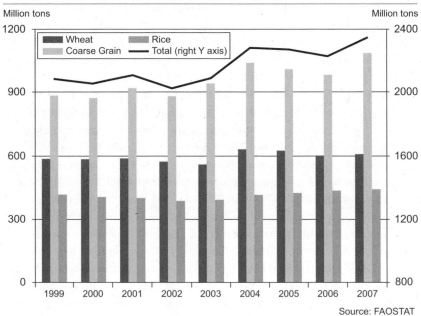

Source: FAOSTAT

Figure 17.2
World Cereal Stocks, 2000–2008

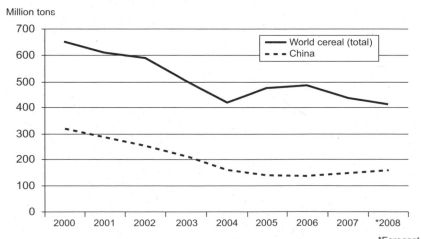

*Forecast
Source: FAOSTAT

and 1980s it declined, finally falling below 2 percent in the 1990s, and it has been below 1 percent for all three for some years in the twenty-first century. This has occurred as the rate of demand growth approached two percent. The slowdown is the result of precipitous declines in public investment in agricultural research and development, and in agricultural development. The principal supplier of international agricultural R&D, the CGIAR, experienced flat or declining real funding for the past twenty years. At the same time. private-sector investment in agricultural research increased but almost exclusively in industrialized countries. This increase is driven by rapid advances in biotechnology and intellectual property protection for plant species, which allow private firms to capture the benefits of their research.

The global governance challenges here focus on how to expand the capacity to produce productivity improvements in developing countries and in the CGIAR. Remember that the rapid increases in public investment in the CGIAR and in developing country agriculture in the 1960s and 1970s contributed to the green revolution. The CGIAR investment provided unrestricted funding to highly focused research institutions that produced among other things, semi-dwarf rice and wheat, which contributed to, for example, the more than doubling of Indian wheat production on less land than was planted in the 1960s.

Figure 17.3

Agricultural Productivity Growth for Rice, Maize and Wheat, 1963–2005

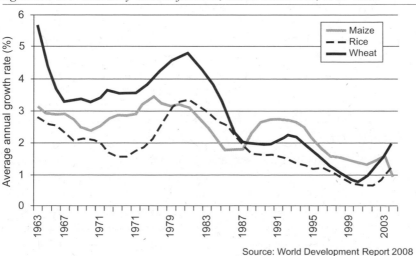

Source: World Development Report 2008

But the CGIAR was a victim of its own success, as donors began to ask it to take on projects that were more about development and less about research. Recent reviews of the CGIAR System have concluded that it has lost some of its focus and is unable to respond collectively and quickly to current challenges. When World Bank president Robert Zoellick proposed doubling global investments in the CGIAR from US$500 million to US$1 billion, many were disappointed at the inability of the CGIAR to respond collectively and rapidly. The CGIAR is a decentralized family of fifteen international research centres, each governed by an independent board of trustees. When it was formed in the early 1970s it was designed to encourage scientific independence and entrepreneurship by individual scientists and centres. The critical question is whether an informal mechanism for collective action invented in the 1970s will be able to respond to a greatly changed world. Not the least of these changes is the biotechnology revolution, which offers much potential but is capital intensive, involving economies of scale. These changes in science, plus the application of intellectual property protection to plant and animal life, have lead to a rapid expansion of private-sector investment. But this expansion has been concentrated in large developed and emerging-market countries, while public investment in developing countries has slowed, leading to concerns about the impact of privatization of agricultural research on food security in developing countries. There are also other new challenges, including global warming, increasing water constraints, and urbanization. And finally, the architecture of aid has changed radically with the multiplication of actors, including thousands of NGOs and new foundations such as The Bill and Melinda Gates Foundation (Kharas 2007).

Restoring higher rates of productivity growth worldwide, and in particular in developing regions, is an enormous challenge. The CGIAR has always been a fragile institution that lacks any formal status, but many would argue it is one of only a few really successful international initiatives. One hopes the reform efforts currently underway do not end up destroying this critical institution.

The Serious Challenge of Reducing Agricultural Protectionism

As noted above, an efficient, open, and fair trading system is clearly an international "public good" of critical importance to all countries. Yet agriculture, particularly in developed countries, remains highly subsidized and protected. The Doha Round of WTO negotiations remains barely alive, on life support after eight years of negotiations. The problem is again attributed to the failure to agree to reduce protection and subsidies to mainly large and well-off farmers in Organization for Economic Co-operation and Development (OECD) countries, and for emerging countries to agree to open their highly protected domestic markets. Why does this matter? It matters in three ways that are important to global food security.

First, high levels of support to domestic producers who sit behind high levels of border protection—for sugar and dairy in most countries—increases domestic production, which either reduces demand for imports or expands export supply. The result either way is depressed world prices, which discourages agricultural production in poor developing countries where the majority of poverty is rural. It also reduces access to promising developed-country markets. The recent international price spike underlines the potential negative impacts of unstable international markets on poor food-importing countries.

Second, it has been conclusively proven in manufactured-products trade that reducing trade barriers contributes to more rapid growth in income and to the reduction of poverty. The record of GATT/WTO in freeing non-agricultural trade since World War II is a remarkable international governance success of the postwar period. Agricultural trade could similarly contribute if it liberalized, but rich-country farm interests are powerful and have been successful in sustaining highly protective domestic farm support programs. The most recent WTO suspension was a stand-off between rich countries and emerging developing countries like India over how much rich countries will reduce domestic support and open borders versus how much traditionally closed economies will be opened to trade. The global economic meltdown in the last half of 2008 and 2009 is so severe that it has kept the Doha Round alive, because there is real concern that national reaction to the global recession will be increased protectionism. Freeing agricultural trade could have a positive effect in pulling the world out of recession.

Third, both trade theory and empirical evidence show that the more open a world market, and the more countries that participate in it, the more stable it will be. This is because protectionism allows countries to export their shocks (i.e., production shortfalls) and protect themselves from sharing in needed global adjustments. As more countries close their borders, random shocks such as droughts are borne increasingly by those few that remain open. Thus, freely functioning world agricultural markets are clearly in the interest of developing countries. The world's experience with the rapid run-up and crash of commodity prices in 2007–09 should alert us to the likelihood that more liberalized agricultural markets would have expressed less volatility.

A WTO solution that moves in the direction of more open agricultural and food trade is a large and difficult global governance challenge. It will require bold steps by both rich countries and developing countries to simultaneously commit to a much more liberalized agricultural trade agenda.

Managing National Interfaces with Unstable World Markets

Regardless of progress on the first two challenges, global commodity markets will periodically be subject to shocks. The causes could be many: random natural events (a coincidence of droughts in major producers); global warming, which is likely to increase the number and magnitude of weather events; economic disturbances like those that have enveloped the world in 2008–09; or radical changes in policy by large countries, such as the Soviet Union entering the world wheat markets in 1971–72 in a big way. Because of the thinness of most agricultural markets, small changes in growth rates of supply or demand can cause severe price swings. We have had three such events since 1970 (1972–74, 1996–97, and 2007–08). Given that many countries, and in fact whole regions, are increasingly dependent on world markets for basic food supplies, rapidly rising world prices can have devastating impacts on poverty and hunger.

Among world regions, the Middle East and North Africa Region (MENA) is the most vulnerable because it has been food deficit for many years and that deficit is rising. Table 17.1 shows the import dependence of selected countries. Rapidly rising food prices, dwindling grain stocks, and mounting concerns about physical shortages of food have raised issues of food security to the top of many national agendas. In the short run, ameliorating the impacts of rising food costs on the poor is the immediate concern. If people live on US$2 or less per day and spend 50–60 percent of their income on basic food, a doubling of food prices could drastically reduce or eliminate all other spending and still reduce food purchasing power. Better policies are necessary to mange price and quantity risk in world markets as are more efficient and effective ways to provide food safety nets. Safety nets are addressed in the next section.

Overall, what is required is a comprehensive food strategy that incorporates dealing with international markets as a permanent feature. This is necessary because, for an increasing number of countries, food self-sufficiency is unlikely to be physically possible and most certainly is economically indefensible. As recent events have proven, international markets can exhibit significant price escalation and instability and raise the possibility of short supplies. Therefore, if a country is faced with permanently importing a significant and growing share of food needs, how can the country mange that risk and instability? Fortunately there are options. One is for countries to acquire stocks in periods of regular or low prices as a hedge against future price increases. However, there are two difficulties with this approach. First, holding stocks is expensive, especially for small countries. Second, agricultural products deteriorate over time, which requires constant rotation of stocks. The first of these could be partially overcome by regional cooperation on stock holding, but the second is a

Table 17.1
Cereal Balances in Selected MENA Countries, 2005

	Production 1000 mt	Net Imports 1000 mt	Consumption 1000 mt	Trade Dependence imports/con %
Algeria	3527	8249	11776	70.1
Egypt	22405	12414	34819	35.6
Iran	21906	4615	36521	17.4
Iraq	3701	3415	7116	47.9
Jordan	1024	1892	2916	64.9
Lebanon	1774	736	2510	29.3
Libya	234	1214	1448	83.8
Morocco	4283	5018	9301	53.9
Saudi Arabia	2999	8204	11203	73.2
Syria	5631	2383	8014	29.7
Tunisia	2132	2437	4569	53.3
Yemen	496	1548	2044	75.7

Source: von Braun et al. 2008

permanent challenge. Joachim von Braun from the International Food Research Policy Institute proposes the need for a new global grain reserve policy while recognizing that past attempts have not been successful (see Table 17.2). However, there are options for holding physical stocks.

Table 17.2
Experience with Global Reserves

A new "global coordinated grain reserve policy" is needed

Past arrangements:
1950s: Global Emergency Food Reserve proposal by FAO Council
1975: Int'l Grain Reserve proposal by US Congress delegation
1976: Int'l Emergency Food Reserve (IEFR) created pending 1975 negotiations
1980s: Proposals to strengthen IEFR were not approved
1990s: EU surpluses
2000s: None

Source: von Braun et al., IFPRI, April 2008

There are a range of options for price risk management. The World Bank Framework Document for a Global Food Crisis Response Program (GFCRP) describes the possibilities in detail. They first discuss utilization of commodity futures markets:

> There are two main approaches to hedging price risk within a commodity chain. **Financial hedging products**, such as futures and options, are typically traded on established commodity exchanges. They are generally used to mitigate short-term price risks, i.e. 3–8 months forward.… Option contracts, which function as a sort of "price insurance" mechanism, are more easily accessible by developing country clients since there is no credit risk carried by the provider if the premium is paid up front. In Tanzania, Costa Rica, and El Salvador, commodity market actors have used option contracts to help manage the risk of short-term price volatility and work on this instrument is ongoing. There are two types of option contracts: i) **Put options** create a price floor, and thus provide protection (i.e. for producers, exporters) against the risk of prices falling, ii) **Call options** create a price ceiling, and thus provide protection (i.e., for consumers, importers) against the risk of prices rising. On the financial market, a call option is based on underlying futures contracts. In the physical market, a call option is based on delivery of physical stocks. (World Bank 2008a, 74; emphasis in original)

A second approach would be customized risk management instruments using physical trading products:

> Given the challenges associated with accessing futures markets directly for many developing country clients, more appropriate risk management instruments can be customized using physical trading products, i.e., forward contracts, minimum-price guarantee contracts, and physical options. These transactions are called **OTC "over-the-counter"** transactions because they are customized to client needs. Physical call options have been used by the Government of Malawi and are currently being evaluated for Haiti. The instrument can provide a country with upside price protection through a contract that would guarantee future delivery of rice at a price no higher than the pre-agreed ceiling price. If prices move higher during the time period of the contract, supply is assured at the pre-agreed ceiling price. If prices move down during the time period of the contract, the option has no value and the country can source rice through other channels, at the lower cost. (ibid., 75)

These options present two serious challenges for developing countries. First, global commodity markets are complex and fast moving. Access to full, current global information and the capacity to act fast are necessary conditions for playing. Furthermore, the financial instruments in futures and options markets are complicated and involve taking positions that have high levels of financial risk.

Second, these markets require instant access to financial resources. For a small developing country in the middle of Africa, these challenges could be prohibitive. Thus, the challenge is for global entities like the international financial institutions (e.g., the World Bank) is to work with regions and countries to develop effective and functional mechanisms of international price risk management.

Food Safety Nets—National and International

Food insecurity arises for many other reasons beyond international market instability, however. It can arise from high levels of domestic poverty, shortfalls in domestic production, persistent regional droughts (e.g., Horn of Africa and Southern Africa in recent years) and national and regional conflicts that disrupt production and national food-distribution systems.

Many countries provide general food safety nets by subsidizing the prices of, or physically providing, basic staples such as bread, flour, sugar, and vegetable oil. If these subsidies are not targeted, they can become major costs, which are greatly escalated by rising world prices if the countries are importers. Again using MENA as an example, subsidies in some countries exceed 2 percent of GDP (Figure 17.4) and up to 8 percent of government expenditures (Figure 17.5). The challenge for these countries has to be to develop more efficient and effective targeted food safety nets, and the international challenge is to assist countries in that task (World Bank 2008b).

Figure 17.4

Food Subsidies in Selected MENA Countries, 2007 (Share of GDP)

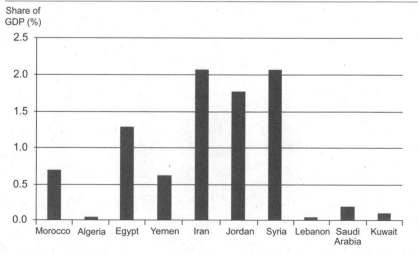

Source: World Bank

The international community can help countries cope with weather risk. Attempts after the World Food Summit in 1996 to develop a Food Insecurity and Vulnerability Information Mapping System were not fully successful but are clearly needed. Recently the World Bank has proposed a mechanism to manage weather risk. Their GFCRP Program proposes a third type of risk management—Early Warning and Weather Risk Management for Crop Production. It would operate by strengthening national crop surveillance and early warning systems and by using an index-based weather risk-management contract. World Bank president Robert Zoellick proposed to the 2008 Rome Food Summit a ten-point plan that included the use of weather derivatives (Zoellick 2008). Designing a country-specific risk-management strategy is clearly an important potential role for agencies such as the World Bank.

Finally, there is a well-established international safety net mechanism for when food emergencies and famine arise: the World Food Programme, which has in the past two years become the world's largest handler of food aid. This organization provides the last international governance defense against food insecurity by delivering food supplies directly, or in cooperation with NGOs, to hungry people. The WFP works in cooperation with bilateral food aid programs such as the US Food for Peace Program (PL 480) and the European Union Food Aid program. Everyone agrees that it is preferable for aid agencies

Figure 17.5

Food Subsidies in Selected MENA Countries, 2007
(Share of Government Expenditure)

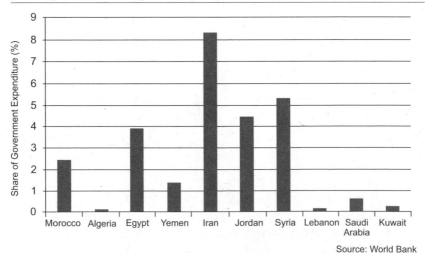

Source: World Bank

to provide cash support for the most efficient and timely purchase of needed food by recipient countries and WFP, yet the elements of PL 480 still insist on providing surplus US commodities. Thus, moving further toward a cash-based system seems desirable. Sadly, much of WFP's continuing efforts are devoted to feeding people displaced by conflict.

Conclusion

This chapter has identified four interrelated global challenges to food security: agricultural productivity improvements, liberalized international agricultural trade policies, the matching of national interfaces with unstable world food markets, and the installation of both national and international food safety nets. Of these, the most critical one is long-term productivity growth, because unless the world is able increase the rate of productivity growth, the future of global food security is at risk. This is not to say the others are not important, because they are. Nonetheless, we can probably survive without a WTO agreement, but we cannot if we do not learn how to produce more with less.

But there is a nagging concern. Those who remember the aftermath of the 1972–74 price run-ups, which was larger than the one we have just experienced, recall that the global community vowed that it would never neglect agriculture again. It was going to invest in research and technology development, pursue agricultural development, and reduce rural poverty, saying that the world must never again be lulled into a false feeling of security by declining real food prices. Further commitments were made for trade reform that let world markets work freely. But by the early 1980s, the world fell victim again to low food prices, surpluses, and agricultural protectionism, and by the late 1980s agricultural development and agricultural research were low on most agendas.

Most grain prices, which peaked up almost 300 percent from where they started in early 2007, are now back down to near the levels they were before the run-up. Given these falls, how long will doubling the CGIAR's budget or increasing investment in regional food security remain high on the agenda of aid agencies, both bilateral and multilateral? Or will it be "how quickly we forget" again? Will it be another twenty-five years before the World Bank publishes another *World Development Report* on the theme of agriculture? We must hope not, because by then there will be almost two billion more people to feed.

Works Cited

Evans, Lloyd (1998). *Feeding the Ten Billion: Plants and Population Growth*. Cambridge: Cambridge University Press.

Food and Agriculture Organization of United Nations (2003). "Trade Reforms and Food Security—Conceptualizing the Linkages." Rome. http://www.fao.org/docrep/005/y4671e/y4671e00.htm.

Kharas, Homi (2007). "The New Reality of Aid." Presented at the *Brookings Blum Round Table*. Washington, DC: Brookings Institution. 1 August.

von Braun, Joachim, Akhter Ahmed, Kwadwo Asenso-Okyere, Shenggen Fan, Ashok Gulati, John Hoddinott, Rajul Pandya-Lorch, Mark W. Rosegrant, Marie Ruel, Maximo Torero, Teunis van Rheenen, and Klaus von Grebmer (2008). "High Food Prices: The What, Who, and How of Proposed Policy Actions." Policy Brief. Washington, DC: International Food Policy Research Institute. May.

World Bank (2007). "Agriculture for Development: World Development Report 2008." Washington, DC: World Bank.

——— (2008a). "Framework Document for Proposed Loans, Credits, and Grants in the Amount of US$ 1.2 Billion Equivalent for a Global Food Crisis Response Program (GFCRP)." Washington, DC. 26 June.

——— (2008b). "Global Financial Crisis: Responding Today, Securing Tomorrow." Background Paper prepared for the G20 Summit on Financial Markets and the World Economy. Washington, DC. 15 November.

Zoellick, Robert (2008). "A 10-point Plan for Tackling the Food Crisis." *Financial Times*. 29 May.

NOTES ON
CONTRIBUTORS

Noora-Lisa Aberman is a Program Analyst and Communications Specialist in the Poverty, Health and Nutrition Division (formerly the Food Consumption and Nutrition Division) of the International Food Policy Research Institute in Washington, DC.

Jennifer Clapp is a CIGI Chair in International Governance and a Professor in the Faculty of Environment at the University of Waterloo. Her recent books include *Paths to a Green World: The Political Economy of the Global Environment* (co-authored with Peter Dauvergne, MIT Press, 2005) and *Corporate Power in Global Agri-Food Governance* (co-edited with Doris Fuchs, MIT Press, 2009). She is also co-editor of the journal *Global Environmental Politics* (MIT Press).

C. Stuart Clark is the Senior Policy Advisor at the Canadian Foodgrains Bank, a coalition of Canadian church-related development organizations. He is also the Chair of the Trans-Atlantic NGO Food Aid Policy Dialogue (TAFAD), a consortium on European and North American NGOs dedicated to the reform of the international food aid regime.

Marc J. Cohen is a Humanitarian Policy Researcher at Oxfam America in Washington, DC, and professorial lecturer in international development at the Paul H. Nitze School of Advanced International Studies, Johns Hopkins University. He previously was a research fellow in the Food Consumption and Nutrition Division of the International Food Policy Research Institute (IFPRI), and was the lead author of the 2008 IFPRI-FAO study *Impact of Climate Change and Bioenergy on Nutrition.*

Kimberly Ann Elliott is a Senior Fellow with the Center for Global Development (CGD). Her most recent books include *Delivering on Doha: Farm Trade and the Poor*, which was co-published by CGD and the Peterson Institute for International Economics in 2006, and *Economic Sanctions Reconsidered* (with Gary Hufbauer and Jeffrey Schott, 3rd ed., 2007).

Daniel Gustafson is the Director of the FAO Liaison Office for North America. He has worked for the past thirty years on agricultural and rural development in Latin America, Africa, and Asia as well as in the United States. Previously, he was Program Director of the International Development Management Center at the University of Maryland.

Raymond Hopkins has taught at Swarthmore College since 1967. He is the author or co-author of six books and over sixty articles. He has been a consultant to the State Department, the Food and Agriculture Organization of the United Nations, the World Food Programme, the Agency for International Development, the Canadian International Development Association, the International Food Policy Research Institute, and the World Bank.

Sue Horton is a Professor of Economics and Associate Provost Graduate Studies at the University of Waterloo. She has worked in over twenty developing countries and has consulted for the World Bank, the Asian Development Bank, several UN agencies, and the International Development Research Centre. She served as the Associate Dean of the Faculty of Arts and Science, Chair of the Department of Social Science and interim Dean at the University of Toronto at Scarborough, and as Vice-President Academic at Wilfrid Laurier University.

Julie Howard is Executive Director of the Partnership to Cut Hunger and Poverty in Africa, an independent nonprofit coalition dedicated to increasing the level and effectiveness of US assistance and private investment in Africa through research, dialogue, and advocacy. She also serves as an Adjunct Assistant Professor of Development at Michigan State University.

Marcia Ishii-Eiteman is a senior scientist and director of Sustainable Food Systems at Pesticide Action Network North America. Previously she worked fifteen years in sustainable agriculture, women's health, and participatory rural development projects in Asia and Africa. Ishii-Eiteman holds a PhD in Ecology from Cornell University and has written extensively on the ecological, social, and political dimensions of food and agriculture. She was a lead author of the UN-sponsored International Assessment of Agricultural Knowledge, Science and Technology of Development (IAASTD).

Gawain Kripke is the Director of Policy and Research for Oxfam America, based in Washington, DC. Prior to that, he served as a Senior Policy Advisor on Oxfam's Make Trade Fair campaign. He is author of numerous opinion pieces and briefing papers on trade and development issues. Before to joining Oxfam, he served as Director of Economic Programs for the environmental organization Friends of the Earth.

John Markie is an independent consultant. He was secretary to the Governing Bodies for the reform of the Food and Agriculture Organization of the United Nations. Prior to that, he served as Head of Evaluation and managed the independent evaluation of FAO. He was also active in the inter-agency United Nations Evaluation Group.

Alex McCalla is Professor Emeritus of Agricultural Economics at the University of California, Davis, where he served as Dean of the College of Agricultural and Environmental Sciences, Associate Director of the California Agricultural Experiment Station, and founding Dean of the Graduate School of Management. After retiring from Davis, he directed the Agriculture and Natural Resources Department at the World Bank, chaired the Technical Advisory Committee of the Consultative Group on International Agricultural Research, and was a founding member and co-convenor of the International Agricultural Trade Research Consortium.

Anuradha Mittal is Executive Director of the Oakland Institute, an independent policy think-tank in Oakland, California. She is an expert on trade, development, human rights, and agriculture issues. Mittal is the author and editor of numerous articles and books, including *Voices from Aftrica: African Farmers and Environmentalists Speak against a New Green Revolution in Africa,* and *2008 Food Price Crisis: Rethinking Food Security Policies.* Previously, she was co-director of Food First/Institute for Food and Development Policy.

Frederic Mousseau is Policy Adviser for Oxfam Great Britain. Over the last fifteen years he has worked with international organizations, including Action Against Hunger, Doctors Without Borders, and Oxfam, designing and overseeing food security intiatives in more than twenty countries as well as studies of food crisies and policies. He is author of numerous articles and reports on world hunger and food policies.

Mark Redwood is Program Leader of the Urban Poverty and Environment section of the International Development Research Centre in Ottawa, Canada. He has published numerous articles on wastewater use for agriculture, and his most recent book is *Agriculture in Urban Planning: Generating Livelihood and Food Security* (IDRC 2008).

Emmy Simmons is an independent consultant on international development issues. She is Co-chair of the Roundtable on Science and Technology for Sustainability at the U.S. National Academies of Science and leads a Roundtable working group on Partnerships for Sustainability. Previously, she had a long career and held a number of positions at the US Agency for International Development (USAID), including Assistant Administrator for Economic Growth, Agriculture and Trade.

Cristina Tirado is investigating adaptation strategies to climate change for food security at the School of Public Health of the University of California Los Angeles, and she consults for FAO, WHO, and other international organizations in this field. She has been the WHO Food Safety Regional Adviser for Europe. She participated in the first WHO/EC project on Climate Change Adaptation Strategies for Human Health and was contributing author of the Health Chapter of the 4th Intergovernmental Panel on Climate Change Assessment Report.

Brian Thompson is a senior nutrition officer in the Nutrition and Consumer Protection Division of the Food and Agriculture Organization of the United Nations, based in Rome, Italy, and was co-author of the 2008 FAO-IFPRI study *Impact of Climate Change and Bioenergy on Nutrition*.

Tony Weis is an Assistant Professor in the Department of Geography at the University of Western Ontario, Canada. He is author of *The Global Food Economy: The Battle for the Future of Farming* (Zed Books, 2007) as well as numerous articles on the global food system, agriculture, and environment.

Noah Zerbe is an Associate Professor of Politics at Humboldt State University, California. He is the author of *Agricultural Biotechnology Reconsidered: Western Narratives, African Alternatives* (Africa World Press, 2004) as well as numerous articles on the political economy of agricultural biotechnology.

INDEX

Books in the Studies in International Governance Series

Alan S. Alexandroff, editor
Can the World Be Governed? Possibilities for Effective Multilateralism / 2008 /
vi + 438 pp. / ISBN: 978-1-55458-041-5

Hany Besada, editor
From Civil Strife to Peace Building: Examining Private Sector Involvement in West African Reconstruction / forthcoming 2009 / ISBN: 978-55458-052-1

Jennifer Clapp and Marc J. Cohen, editors
The Global Food Crisis: Governance Challenges and Opportunities / 2009 /
xviii + 270 pp. / ISBN: 978-1-55458-192-4

Andrew F. Cooper and Agata Antkiewicz, editors
Emerging Powers in Global Governance: Lessons from the Heiligendamm Process /
2008 / xxii + 370 pp. / ISBN: 978-1-55458-057-6

Jeremy de Beer, editor
Implementing WIPO's Development Agenda / 2009 / xvi + 188 pp. /
ISBN: 978-1-55458-154-2

Geoffrey Hayes and Mark Sedra, editors
Afghanistan: Transition under Threat / 2008 / xxxiv + 314 pp. /
ISBN-13: 978-1-55458-011-8 / ISBN-10: 1-55458-011-1

Paul Heinbecker and Patricia Goff, editors
Irrelevant or Indispensable? The United Nations in the 21st Century / 2005 / xii +
196 pp. / ISBN 0-88920-493-4

Paul Heinbecker and Bessma Momani, editors
Canada and the Middle East: In Theory and Practice / 2007 / ix + 232 pp. /
ISBN-13: 978-1-55458-024-8 / ISBN-10: 1-55458-024-2

Yasmine Shamsie and Andrew S. Thompson, editors
Haiti: Hope for a Fragile State / 2006 / xvi + 131 pp. / ISBN-13: 978-0-88920-510-9 /
ISBN-10: 0-88920-510-8

Debra P. Steger, editor
Redesigning the World Trade Organization for the Twenty-first Century / forthcoming
2009 / ISBN: 978-1-55458-156-6

James W. St.G. Walker and Andrew S. Thompson, editors
Critical Mass: The Emergence of Global Civil Society / 2008 / xxviii + 302 /
ISBN-13: 978-1-55458-022-4 / ISBN-10: 1-55458-022-6

Jennifer Welsh and Ngaire Woods, editors
Exporting Good Governance: Temptations and Challenges in Canada's Aid Program /
2007 / xx + 343 pp. / ISBN-13: 978-1-55458-029-3 / ISBN-10: 1-55458-029-3